THE DOs

Dr. Andrew Taylor Still, 1828–1917

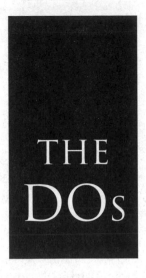

THE DOs

OSTEOPATHIC MEDICINE IN AMERICA

Second Edition

NORMAN GEVITZ

THE JOHNS HOPKINS UNIVERSITY PRESS
Baltimore & London

The Johns Hopkins University Press
2715 North Charles Street
Baltimore, Maryland 21218-4363
www.press.jhu.edu

Library of Congress Cataloging-in-Publication Data

Gevitz, Norman.
The DOs : osteopathic medicine in America / Norman Gevitz.–2nd ed.
p. ; cm.
Rev ed. of: The D.O.'s. c1982
Includes bibliographical references and index.
ISBN 0-8018-7833-0 (alk. paper) — ISBN 0-8018-7834-9 (pbk. : alk. paper)
1. Osteopathic medicine—United States—History.
[DNLM: 1. Osteopathic Medicine—history—United States.
WB 940 G396d 2004]   I. Gevitz, Norman. D.O.'s.   II. Title.
RZ325.U6G48 2004
615.5'33'0973—dc21
2003012874

A catalog record for this book is available from the British Library.

Frontispiece courtesy of the Still National Osteopathic Museum, Kirksville, Missouri.

For
Kathryn Gevitz

CONTENTS

PREFACE & ACKNOWLEDGMENTS

I first became aware of the existence of the osteopathic medical profession during the summer of 1974. I was meeting my friend David, who was soon to graduate with his MD degree from a Chicago medical school. We were going to play tennis. The court we had reserved was still in use, and while we waited for it we got into a conversation in which I brought up the subject of "occupational role duplication." I was a sociology graduate student at the University of Chicago and was interested in the phenomenon in which one profession offered a range of services to the public that overlapped with services provided by one or more other professions. I was curious about the political and legal aspects of how groups carved out and maintained role boundaries. I knew a little about health care services, particularly that several professional groups competed with each other, causing considerable conflict between them in regard to what should constitute their respective domains and scopes of practice. I thought my medical student friend could offer a helpful perspective based on his background.

My first question to David concerned the struggle between ophthalmologists and optometrists. The former went to medical school and claimed the entire clinical field regarding eye care. Optometrists, who were trained in their own schools, were licensed, entitled to use the title *Doctor*, and were doing refractions like the ophthalmologists. Optometrists, as I recall, were then seeking changes in the law to extend their scope of practice to allow them to use drugs to dilate the pupils in order to diagnose glaucoma and other eye diseases. My friend told me what he knew about optometry and why the Illinois State Medical Society was opposed to any expansion in optometric practice rights. I then asked him about the shifting practice boundaries between oral surgeons (trained in dental schools) and MDs who specialized in head and neck surgery. After he gave me an informative answer, I inquired about professional conflicts between podiatrists and orthopedists. Podiatrists, who had long confined their interventions to diseases of the foot up to the ankle, wanted to change their scope of practice to include the knee, and some wished to extend their involvement to the thigh (a sort of professional gangrene, I thought). Once again, my friend confidently and competently answered my questions about the boundaries and divisions between these two groups.

Before I could ask my next question, David looked at me warily and said, "Please don't ask me to tell you what the difference is between an MD and a DO!" After a pause, I had to inquire, "What's a DO?" He replied, "An osteopath." I looked at him blankly and asked, "What's an osteopath?" My friend rolled his eyes, thought a little, and finally said, "Well, DOs are licensed to practice medicine like MDs. They treat the entire body. However, they have their own colleges, hospitals, associations, specialty boards, and their own journals. They use medicines and surgery, but they also employ spinal manipulation. They are like us but different. That's all I know and you would have to ask somebody else for more." At that point, the tennis court came open, we played our match, and the subject was dropped.

Being a graduate student I thought (incorrectly) that I was quite worldly, so it caught me by surprise that I had never heard of osteopathic medicine. The next day, simply curious, I strode into Regenstein Library on the University of Chicago campus seeking further information. I found nothing during that particular visit. Back in the sociology department I asked my graduate student friends and a few faculty members whether they had heard of osteopathy, and none of them responded affirmatively. On the way home that evening, I came to the conclusion that my friend had played a little joke on me—there was no such thing as a DO. Nevertheless, when I got home I thought of one last resource. I went to my copy of the *Encyclopedia Americana*, which was then at least twenty years old, and to my surprise I did find a heading "osteopathy." While brief, the entry essentially confirmed the outlines of what my friend had said about the profession. It noted that there were then six schools of osteopathic medicine, one of which was in Chicago. Turning to the phone book, I found the address and was amazed to find that this school—the Chicago College of Osteopathic Medicine (CCOM)—was located only five blocks north of the University of Chicago. I could not comprehend how, despite the close proximity of the osteopathic school, nobody whom I had talked to on my campus was aware of DOs or the osteopathic profession.

By now I was intrigued, and the next day I made the short trek north and headed for the CCOM library. There I found a complete run of the *Journal of the American Osteopathic Association (JAOA)*. Picking up the most recent issues, I sat at a long table and read through them. Although I was not medically trained, I could observe that in many respects the *JAOA* appeared to be very much like MD medical journals on the shelf. The articles appeared in conventional medical format, and the standard course of treatment recommended by the authors involved the use of drugs and surgery. Most telling to my untrained eye, all of the most recent issues of the *JAOA* were full of ads by pharmaceutical manufacturers. There were, however, differences from the MD journals. I noticed that some of the articles discussed palpatory diagnosis of the spine and manipulative treatment in the overall management of patient problems, and a few articles were solely devoted to the benefits and applications of manipulative intervention.

I next turned to the very first issues of the *JAOA*, which had been published in the year 1901. Perusing these initial volumes, I was immediately struck by how different the articles were. All were focused on palpatory diagnosis and manipulative therapy and contained broad discussions of osteopathic principles and practices. According to the authors, the various diseases treated had one principal cause—displacements of vertebrae along the spinal column. These "lesions," it was believed, interfered with nerve and blood supply and were responsible for disordered physiology throughout the body. There was no mention of drugs except in the most disparaging or dismissive way. This transformation in beliefs and practices by DOs over three-quarters of a century captivated me. How, I asked, did this profession originate? How did it evolve from producing limited practitioners to fully licensed physicians and surgeons? How was it able to survive and grow despite what I was learning had been unremitting opposition of the MDs? Why did so few people know what a DO was? And finally, how had changes in the profession over recent decades impacted its professional autonomy?

Given the paucity of secondary sources and my growing fascination with the subject, I decided to make this my doctoral dissertation, under the guidance of the medical sociologist Odin W. Anderson, PhD, and the medical historian Lester S. King, MD. In undertaking this project I was most fortunate that, in addition to the nearby osteopathic college, the headquarters of the American Osteopathic Association (AOA) was located in Chicago. The executive director of the AOA, Edward Crowell, DO, permitted me wide access to its extensive library and archives, where I spent most of my research time reading complete runs of several osteopathic journals and wading through school catalogs, college and hospital accreditation surveys, AOA Committee reports and other materials. It took five years to complete the dissertation and another two years before I saw the manuscript through to publication in 1982 by the Johns Hopkins University Press.

With the appearance of *The DOs*, my research shifted to other areas within medical history, though from time to time I would return to look at the osteopathic profession. For twelve years I taught MD students at the University of Illinois College of Medicine at Chicago, where I became professor of the history of medicine and head of the medical humanities program. In 1997, I left to accept an offer from Barbara Ross-Lee, DO, then dean of the Ohio University College of Osteopathic Medicine, to become professor and chair of their new Department of Social Medicine. This position allowed me to teach osteopathic medical students and provided me with an unparalleled opportunity to critically examine from the inside the strengths and weaknesses of the osteopathic profession.

In 2000, I decided to undertake a second edition of *The DOs*. Although the book has been in print continuously since its initial publication, it had with each passing year become more dated. For the second edition I have added material, improved the narrative, and corrected small errors in the first nine chapters of the original edition. I have also added two new chapters, "In a Sea of Change" and "The Challenge of Distinctiveness," which have brought the history of the profession up to the beginning of the millennium.

Much has happened in the more than twenty years since *The DOs* initially appeared. Osteopathic medicine became the fastest growing segment of the physician and surgeon population in the United States. In 1982 approximately 19,000 DO physicians and surgeons were in active practice. In 2003 the total reached 50,000. Currently, more than 60 million patient visits are made to the offices of DOs. When I began my research back in 1974, there were nine osteopathic colleges. At the time of this writing, twenty schools exist, and plans for other colleges are in development. Yet, despite its remarkable numerical expansion, osteopathic medicine faces difficult problems: the decline in use of distinctive osteopathic manipulative procedures, a too narrow range of funding sources for its educational institutions, inadequate public visibility, and the growing assimilation of many DOs into the medical mainstream. As a consequence, an overriding question that osteopathic medicine needs to answer is whether it can carve out for itself a continuing and distinctive role in the American health care system and maintain its professional autonomy.

IN THE FIRST EDITION, I thanked the following individuals for their help: Charles Bidwell, PhD; Anne Crowley, PhD; William Cummings, PhD; J. Stedman Denslow, DO; Jane Denslow; Michael Doody; Morris Fishbein, MD; Melanie Gevitz; Murray Goldstein, DO, MPH; Philip Greenman, DO; Leonard Heffel; Elliot Lee Hix, PhD; Morris Janowitz, PhD; Robert Kappler, DO; Robert Kistner, DO, MD; Irvin Korr, PhD; Norman Larson, DO; Richard MacBain, DO; Nicholas S. Nicholas, DO; George Northup, DO; Deborah Otis; David Oxman, MD; James Paster, PhD; Michael Patterson, PhD; Barbara Peterson; Anders Richter; Linda Stellato; Edward G. Stiles, DO; Robert Thompson, EdD; Jacqueline Wehmueller; and Linda Westerfield.

To this group I would like to add the following people who assisted me in a variety of ways to produce this second edition: William Anderson, DO; Steve Andes, PhD; Barbara Barzansky, PhD; Kathryn Bazylewicz; Jack Brose, DO; Norbert Budde, PhD; Boyd Buser, DO; Tim Creamer; John Crosby, JD; Bruce Dubin, DO, JD; Judith Edinger; Michael Fitzgerald; Deborah Heath, DO; Jennifer Jacobs; William Johnston, DO; Mitch Kasovac, DO; Cathy Kearns; Albert Kelso, PhD; Donald Krpan, DO; Howard Levine, DO; Cheri McFee; Christopher Meyer, DO; Steve Noone; Eugene Oliveri, DO; George Reuther; J. Jerry Rodos, DO; Susan Sagusti; Phil Saigh; Michael Seffinger, DO; Allen Singer, PhD; Ida Sorci; Jane Stark; Joseph Vorro, PhD; Anne Whitmore; James Whorton, PhD; Jacqueline Wolf, PhD; Douglas Wood, DO, PhD; and James Zini, DO.

Research for both editions of this book was conducted at the American Osteopathic Association in Chicago; the American Association of Colleges of Osteopathic Medicine in Chevy Chase, Maryland; the American Osteopathic Hospital Association in Chevy Chase; the American Medical Association in Chicago; Missouri State Historical Society in Columbia; the Kirksville College of Osteopathic Medicine in Kirksville, Missouri; the John Crerar Library in Chicago; the National Library of Medicine in Bethesda, Maryland; the Chicago College of Osteopathic Medicine; the

University of Michigan Libraries in Ann Arbor; the University of Chicago Libraries; Alden Library of Ohio University in Athens; the Learning Resource Center of the Ohio University College of Osteopathic Medicine in Athens; and the Archives and Special Collections of the University of California-Irvine. I wish to thank countless helpful individuals at these places.

Travel to several locations to conduct research for this second edition was aided by a grant from the Office of Research of the Ohio University College of Osteopathic Medicine, for which I am most grateful.

THE DOs

ANDREW TAYLOR STILL

Like other medical prophets or revolutionaries, the founder of osteopathy, Andrew Taylor Still, sought recognition as a completely original thinker. In his autobiography, Still maintained that the precepts of his approach came to him in a single moment of inspiration, that no contemporary belief system or practice significantly influenced his theory that most diseases were directly or indirectly caused by vertebral displacements and that elimination of the latter through spinal manipulation would remove symptoms of pathology elsewhere in the body. Although his followers and others later modified this unlikely interpretation of the profession's origin, they did not go far enough in identifying the intellectual currents that had shaped his thought.

Still was born on August 6, 1828, in Jonesville, Virginia, the third of nine children. His father Abram, had served as a Methodist preacher but at the time of Andrew's birth was supporting his wife, Martha, and their offspring by farming and practicing medicine. However, in 1834, when Andrew was six, Abram once again heard the call, sold his land for the then considerable sum of $900, and moved his family to New Market, Tennessee, where he had received an appointment to preach.[1]

In the early decades of the nineteenth century, the Methodist Church in the United States sent its ministers to follow the steady westward march of the population, making each one responsible for a large geographical area known as a circuit. Often, after spending a few years in one location, a preacher would be transferred, so that he might face a new challenge elsewhere.[2]

Of these country clerics, Horace Bushnell wryly noted that they were "admirably adapted, as regards their mode of action to the new west—a kind of light artillery that God has organized to pursue and overtake the fugitives that flee into the wilderness from his presence. The new settler reaches the ground to be occupied, and by the next week, he is likely to find the circuit crossing by his door and to hear the voice of one crying in the wilderness 'The kingdom of God is come nigh unto you.'"[3]

Andrew's first schooling came in Jonesville at the hands of a man named Vandenburgh. "He looked wise while he was resting from his duties," Andrew recalled, "which were to thrash boys and girls, big and little, from 7 a.m. to 6 p.m. with a few lessons in spelling, reading, writing, grammar, and arithmetic . . . [pardoning] our many sins with the 'sparing rod.'" At New Market, Andrew attended classes with his older brothers at an academy called the Holston College. This institution, much to their relief, was conducted by a man "of high culture, a head full of brains, without any trace of brute in his work." In 1837 Reverend Still was transferred to Macon County in northern Missouri, and for two years Andrew's studies were suspended until his father was able to find a regular tutor. From Macon County, the family moved to Schuyler County, and again there was an interruption. But from 1842 through 1848, when he was twenty, it appears, Andrew continuously pursued a formal education.[4]

The family of a circuit rider led an especially rough life on the frontier. There were the periodic moves, and Reverend Still was called away from home on his religious work several times a year for intervals lasting as long as six weeks. The salary provided him by the church was insufficient to provide for his large brood, and Reverend Still had to supplement the preaching income with earnings from his farm and medical practice. As a child Andrew devoted much of his time to chores such as caring for the crops and livestock. He much preferred hunting. On occasion he traveled with his father on ministerial rounds and participated in the camp meetings that the Reverend Still helped lead. At these religious revivals songs were sung, prayers were offered, and conversions were made. The keynote was enthusiasm. In describing this phenomenon, William Sweet, in his history of American Methodism, observed, "The revival in many instances was accompanied by certain peculiar bodily exercises, such as jerking, rolling, barking, dancing, and falling. The falling exercise was the most common, and frequently at these great meetings scores, even hundreds were on the ground, many lying for a considerable length of time either entirely unconscious or semi-conscious. The 'jerks' were also common, though affecting different persons in different ways. Sometimes the head would be affected, twisting it rapidly to the right and left; sometimes it would seize the limbs, sometimes the whole body."[5] In retrospect, it is surprising that these events did not then give Andrew the idea that anatomical displacement was the predisposing cause of most diseases.

In 1851 Abram received yet another appointment, this time to the Kansas Territory as missionary to the Shawnee Indians. For the time being Andrew remained behind, married, and began working a farm of his own. Two years and two children later, however, he and his family joined his parents at the Wakarusa Mission. It was here that Andrew decided on medicine as a career and began to study and practice under the tutelage of his father. When the territory was officially opened up to settlement soon afterwards, the Stills headed for Baldwin, about twenty miles southeast of Lawrence.

This area became a focus of the national debate over slavery. In 1854 Senator Stephen Douglas of Illinois, seeking to open up the frontier to commerce, introduced legislation organizing the land west of Missouri and Iowa as one territory. Southerners opposed the idea because they believed this proposed geopolitical entity would eventually be admitted as a Free State, thereby altering the existing balance of power in Congress. To win southern support, Douglas amended his bill, calling for the creation of two territories—Kansas and Nebraska—and for the repeal of the Missouri Compromise of 1820, which forbade slavery north of the thirty-sixth parallel. Under the Douglas plan citizens of each territory would decide for themselves whether or not they wanted "the peculiar institution." Over considerable northern objection the bill narrowly passed, and President Pierce signed it.[6]

The result was chaos, as settlers representing both sides of the issue poured into eastern Kansas. In November 1854, an election was held to choose a delegate to Congress. Hundreds of proslavery Missourians crossed the border to vote and were successful in selecting one of their own. The next March, they returned and elected a legislature that promptly enacted a slave code. Following this string of setbacks, the abolitionist forces in the area began to rally. Among them was Andrew Still, who became a lieutenant to the movement's leader, James Lane. In 1857 Still was elected to the quasi-legal Free Kansas Legislature, which passed its own set of laws and organized the people to vote down the existing constitution.[7] After three years of continuing political debate, as well as intermittent bloodshed, Kansas was finally admitted to the Union as a Free State just prior to the beginning of the Civil War.

At the outbreak of the national conflict Still enlisted in the northern cause and was assigned to the 9th Kansas Cavalry, Company F, as a hospital steward; he was responsible for the procurement of drugs and other medical supplies. In April 1862, after being released from this service, he returned home, organized his own command, and was commissioned Captain. Later Still was transferred to the 21st Kansas Militia with the rank of Major. In 1864 he saw action in the successful drive against Confederate forces advancing upon Kansas City.[8]

"During the hottest period of the fight," he recalled, "a musketball passed through the lapels of my vest, carrying away a pair of gloves I had stuck in the bosom of it. Another minie-ball passed through the back of my coat, just above the buttons making an entry and exit almost six inches apart. Had the rebels known how close they were to shooting osteopathy, perhaps they would not have been so careless."[9] This battle marked the end of Still's military career. Returning to Baldwin, he resumed his fledgling career as an orthodox physician.

STILL AND MIDCENTURY MEDICAL PRACTICE

American medicine in the 1850s and 1860s was generally characterized by poorly trained practitioners employing harsh therapies to combat disease entities they understood insufficiently. Before the Civil War the great majority of physicians had never attended a medical school; they either had been trained through the apprenticeship system or were engaging in practice without benefit of any formal background.[10]

The apprenticeship, which could last three or more years, afforded the student a pragmatic education. After reading anatomy and physiology with a preceptor, the trainee learned how to diagnose, how to compound and administer drugs, and how to perform common minor surgical procedures. The qualifications of the preceptor were not standardized and instruction was usually poor, given the paucity of adequately trained physicians, mainly those who had received their education abroad. Nevertheless, the system itself was quite popular, providing the teacher with a dependable income and a cheap source of labor from the student, who in turn received the knowledge necessary to practice medicine according to public expectations.[11]

After serving an apprenticeship the student could elect to enroll in one of the growing number of medical colleges springing up in the country. The aim of these institutions was to supplement the training already received with formal lectures and demonstrations. Initially, instruction in these schools was brief, consisting of one term four to six months in length taken in two successive years, the second term being merely a repetition of the subject matter assigned in the first. The quality of education in such colleges was not good. As they existed to make a profit for their stock-holders, their expenditures for equipment, facilities, and instructors were relatively modest.

When A. T. Still began his career in 1854, his medical education consisted of work performed at his father's side and the study of a number of texts in anatomy, physiology, surgery, and *materia medica*.[12] His first patients were the Shawnee. "I soon learned to speak their tongue," he reminisced, "and gave them such drugs as white men used, cured most of the cases I met, and was well received." The Indians also constituted the source of his continuing education in anatomy and pathology, as he made occasional nighttime raids into their burial grounds to disinter corpses for dissection. Still noted that, although his conscience was troubled over this, at least his subjects never complained.[13] During the Civil War, Still may have received further training and experience in treating trauma as well as camp diseases. Andrew maintained that his duties far exceeded those subsumed under his title of hospital steward.[14] Late in life, Still also claimed he had gone to medical school. An article published under his name in the *Ladies Home Journal* but actually penned by his grandnephew, declared that he had attended a medical college in Kansas City prior to the Civil War. After this article appeared, the question of what his formal medical education had included was directly posed to Still. He answered that he had attended the Kansas City college just after the war, in the winter of 1865 and 1866 but had stayed just for one term since he was so disgusted with the training. This answer raises more questions, because the first documented medical school in Kansas City did not open its doors until 1869. No records have been found to establish whether Still, in fact, actually attended.[15]

Medical thought and practice in these years was highly speculative and largely empirical. Most American physicians believed that disease was due to organic decomposition, climate, heredity, and mechanical injuries. The germ theory, which had lost favor in the first half of the century, was just beginning to be revived. More than cause, the practitioner was preoccupied with effects. Many physicians thought of disease as the sum total of symptoms and reasoned that the faster each was removed (the temperature lowered, pulse restored to normal, bowels evacuated, or stomach purged), the more rapid the patient's recovery would be. Those disorders bearing common attributes tended to be treated by similar or identical methods, serving only to encourage the use of panaceas such as bloodletting.

Playing an instrumental role in this trend was Benjamin Rush (1745–1813), a signer of the Declaration of Independence and perhaps the most influential American physician of his time. Rush believed that the basis of all disease was physiological tension, particularly of the veins and arteries. In his treatment of this condition, he found the drawing of blood most effective. In fact, bleeding the patient for an acute illness became his practice and teaching, and in later years he even claimed that often the only equipment the physician needed for house calls was the lancet.[16] The results of moderate bloodletting were dramatic and seemed palliative: a sudden drop in temperature, profuse sweating, and a sense of calm. Some practitioners believed that utmost benefits were achieved when patients were bled to unconsciousness. Although the theory upon which Rush based his practice was discredited shortly after his death, bloodletting, which was conscientiously performed by the majority of orthodox physicians in America, remained a popular treatment for yellow fever, cholera, typhoid, typhus, smallpox, croup, and enteritis until the 1850s.[17]

Another widely employed panacea was calomel, a mercury compound that acted as a powerful cathartic. So popular was this remedy in the mid-nineteenth century that it was commonly referred to as "the Samson of the *materia medica*."[18] Since most physicians felt that in treating internal ailments a cleansing of the system was desirable, calomel was often prescribed and not infrequently administered in conjunction with bloodletting. In large doses it was responsible for some dangerous side effects. As Guenther Risse has noted:

> Within a few days after ingestion, severe stomatitis with excessive salivation appeared. Patients had ulcerated lips, cheeks, and tongue, soreness and inflammation of the gums, plus loosening and frequent loss of teeth. Some unfortunate children died with perforation of their cheeks, bucal gangrene, and osteomyelitis of the maxillary bones. Generally there was a gastric pain associated with vomiting and gastrointestinal cramps after ingestion of the calomel. In some cases bloody diarrhea occurred which was ascribed to the disease. . . . The larger doses were considered to have a so-called sedative effect, no doubt because of the more severe systemic consequences of the mercurial poisoning.[19]

Because the drug would not readily pass from body tissue, several years of even intermittent use would produce a cumulative reaction. Still recalled that when he was about fourteen years old he was salivated with mercury which loosened his teeth eventually making him use a partial set of dentures. "I lived in a day," he recalled, "when people had no more intelligence than to make cinnabar of my jawbone."[20]

In addition to calomel other toxic pharmaceuticals of dubious value, such as arsenic, antimony, tartar emetic, lobelia, strychnine, and belladonna, were generally employed. A small number of truly useful agents were available in this era: quinine for malaria, colchicine for gout, opium for pain, and digitalis for dropsy. But each was utilized in the treatment of a host of other ailments, for which their introduction was either of no assistance or even harmful.[21]

Such symptomatic management was not accepted by all orthodox physicians. As early as 1835 Jacob Bigelow introduced the concept of the "self-limiting disease," which he defined as "one which received limits from its own nature and not from foreign influences; one which after it has obtained [a] foothold in the system, cannot in the present state of our knowledge be eradicated or abridged by art."[22] Through a careful study of the drastic, or what he commonly called "heroic" measures then in use, Bigelow concluded that none significantly improved the patient's chances for recovery. Though the article that announced his beliefs received favorable notices in the medical press, the reaction of many of Bigelow's contemporaries was indifference. Depletive measures continued to be employed.

In 1860 Oliver Wendell Holmes, Sr., MD, declared in frustration, "If the whole *materia medica* as now used could be sunk to the bottom of the sea, it would be all the better for mankind—and all the worse for the fishes."[23] Holmes subsequently became the object of severe criticism, as did others who attacked the prevailing practices. In 1863, when William Hammond, MD, the surgeon general of the United States, issued an order removing both calomel and tartar emetic from the Union Army supply table, the doctors revolted. Hammond was blasted by the medical societies and his directive was never enforced.[24] Though physicians could appreciate the arguments in favor of the concept of self-limiting diseases, it was not practical for them to follow the logic of this approach. Many felt it was their role to act; their patients' expectations were other than to have them sit by passively, simply watching and waiting.

In the early Kansas years malaria was probably the leading cause of adult mortality. Though the benefits of quinine were widely known to the first generation of settlers, the drug was quite expensive and often difficult to secure. Also decimating the population was smallpox. Many Kansans doubted the efficacy of vaccination, which had been popularized by Edward Jenner (1749–1823), and never bothered to submit to the procedure. Some feared that vaccination was dangerous and would only spread the disease. Other scourges for which there were no effective therapies—typhoid fever, pneumonia, scarlet fever, typhus, dysentery, and meningitis—were all frequent visitors to the pioneers' homes.[25]

In treating clients suffering from these and other conditions, Still employed such generally accepted drugs as castor oil, gamboge, aloes, lobelia, quinine, and soap pills. Though he may, as he said, have harbored some doubts about their relative value at the beginning of his practice, this did not stop him from prescribing them.[26] Only when tragedy struck his own household in the spring of 1864 did Still begin seriously to question the practice of regular medicine. "War," he wrote, "had left my family unharmed; but when the dark wings of spinal meningitis hovered over the land, it seemed to select my loved ones for its prey." Following the interdiction against treating one's own close relatives, Still summoned nearby physicians, who took immediate charge. He recalled the scene:

> Day and night they nursed and cared for my sick and administered their most trustworthy remedies, but all to no purpose. The loved ones sank lower and lower. . . . God knows I believed they did what they thought was for the best. They never neglected their patients and they dosed and added to and changed doses, hoping to hit upon that which would defeat the enemy; but it was of no avail. It was when I stood gazing upon three members of my family . . . all dead from the disease of spinal meningitis that I propounded to myself the serious questions "In sickness had God left man in a world of guessing? Guess what is the matter? What to give and guess the result?"[27]

While Still would not abandon orthodox medicine per se for another ten years, this personal loss inspired him to evaluate various alternative systems of practice which had already arisen. "Like Columbus," he said, "I trimmed my sail and launched my craft as an explorer."[28]

ALTERNATIVE MEDICAL SYSTEMS

Several vastly different medical movements arose in America beginning in the nineteenth century.[29] The first significant challenge to orthodox medicine in America was led by Samuel Thomson (1769–1843), a crude, self-educated individual who postulated that all disease was due to the body's inability to maintain its natural heat. As therapy he rejected blood-letting and calomel, employing instead botanical remedies that caused the patient to sweat and vomit. Thomson attacked the legitimacy and integrity of the medical profession on several grounds, arguing that the motive of regular physicians was often to obtain a larger fee by prolonging illness, that formal education was an unnecessary prerequisite to practice, and that licensing laws passed on the grounds of protecting the public against "quacks" were only the means by which one group could monopolize the healing arts. Though they were ridiculed by orthodox doctors, Thomson's attacks appealed to many Americans in the age of Jackson, when the virtues of the common man were extolled and the according of special privilege to anyone was frowned upon.[30]

Thomson, however, was not loathe to obtain his own special privilege, securing a patent on his system of medicine and selling family rights for its use at $20 apiece under a slogan stating that every man could be his own doctor. Mobilized by Thomson into "friendly societies," his followers lobbied intensively in state legislatures against existing licensing laws that restricted medical practice to orthodox physicians. By the 1840s almost all of these statutes had been repealed, amended, or otherwise made ineffective. This meant that anyone could practice medicine practically anywhere in the country without fear of being prosecuted, a situation that lasted for several decades.[31]

A distinctly different and more intellectual threat to the medical establishment was presented by homeopathy, which was adopted by thousands of educated physicians in the United States who had been trained in the orthodox tradition. This movement was originally launched in Germany by Samuel Hahnemann (1755–1843), an erudite university graduate who, like Thomson, opposed the standard remedies then in use. In the 1790s Hahnemann began performing experiments on himself, recording the physiological reactions produced by various drugs. The first drug he tested was cinchona bark, from which quinine is derived. He found that if he ingested it while perfectly healthy, his body would manifest several of the symptoms of malaria. This led him to conclude that the drug best able to cure a given disease was the one that produced most of its symptoms in a well person. Other agents were tested by Hahnemann and his followers, who found the use of such homeopathic or "like cures like" remedies most effective, especially when administered in extremely small amounts.[32]

The homeopaths developed their own comprehensive *materia medica* and offered their system as a substitute for the practices of orthodox doctors, whom they labeled *allopaths*. The allopath, declared Hahnemann, was one who would offer treatments that produced completely opposite effects of the disease when administered in health. In subsequent decades, however, the term *allopath* lost its original signification and became a convenient label used by all alternative medical movements in describing "regular" or "orthodox" physicians.[33]

The rapid growth of homeopathy can be easily understood. Its followers did not administer toxic levels of the standard pharmaceuticals of the day, nor did they employ bloodletting. Thus, patients had only to bear the disease, not the treatment as well. Even Holmes, its arch critic, who would deny any physiological efficacy of infinitesimal doses, would nevertheless note that homeopathy "taught us a lesson of the healing faculty of nature which was needed, and for which many of us have made proper acknowledgments."[34]

Before 1860 most American homeopaths were educated in regular medical colleges and learned the Hahnemannian system following graduation. After the Civil War, practicing homeopaths were increasingly expelled from the American Medical Association and orthodox societies and institutions. As a result, they built their own schools and hospitals. The instruction in these colleges was as complete as that in allopathic schools of the day, the two differing only in the content of their materia medica and their approach to patient care. Based on his analysis, William Rothstein concluded that the facilities, staff, and clinical opportunities available in most homeopathic institutions were equivalent to those found in their orthodox counterparts.[35]

Due partly to the increasing popularity of homeopathy, a schism occurred within the Thomsonian ranks. When the founder wanted to restrict his followers' therapeutic armamentarium to the drugs he used and opposed any formal medical training, several of his more sophisticated supporters could not agree. They wished to experiment with all available botanicals as well as any other agent that held promise. Under the leadership of Wooster Beach (1794–1868), this group arose and opened their own colleges, eventually adopting the name *eclecticism* to describe their desire to shun all restrictive tenets or principles. In actual practice, however, the eclectics, who rejected most drugs of mineral origin and substituted resinous medicinals for the regulars' alkaloid pharmaceuticals, did so purely on dogmatic grounds. As for the schools they established before and after the Civil War, most were academically poor, and the physicians they produced were not as well trained as those graduating from their competitors' colleges.[36]

Whatever their respective strengths or shortcomings, unorthodox practitioners flourished as the century progressed. At their apex, they collectively represented approximately 15 percent of the entire physician population; the homeopaths were located principally in the cities and the eclectics mostly in rural areas.[37] It is difficult to establish how much influence these systems had on the practices of regular physicians, who soon abandoned their heroic measures, but certainly their success in attracting a sizable number of patients gave pause to many in the orthodox ranks.

In studying these reform movements, Still came to the realization that they generally offered a less harmful regimen to their patients than did the regulars. However, he concluded that they were just as empirical and ineffective. "First I tried the ability of drugs as taught and administered by allopathy," he once observed, "then noticed closely the effect from the schools of eclecticism and homeopathy. I concluded all was a conglomerate mess of conjectures and experiments on the ignorant sick man from the crown to the heel. I learned that a king was just as ignorant of the nature of disease as was his coachman, and the coachman no wiser than his dog."[38] The central issue in medicine, he would maintain, was not which drug to use and in what dosage, but whether drugging itself was a scientific form of therapy.

No less important was his moral concern. Being a Methodist, Still abstained from alcohol.[39] After the war he asked himself, If drinking is sinful, should drugs be classified any differently? "I was not long in discovering," he reflected, "that we had habits, customs, and traditions no better than slavery in its worst days and far more tyrannical." For this he laid the blame on the physician, proclaiming the cause to be ignorance in "our schools of medicine." He observed, "I found that he who gave the first persuasive dose was also an example of the same habit of dosing and drinking himself, and was a staggering form of humanity wound hopelessly tight in the serpent's coil."[40] Increasingly convinced that internal medication of any kind was immoral as well as invalid, Still would continue his explorations in a different direction.

DRUGLESS SUBSTITUTES

This was an era in which a number of drugless systems appeared and gained some degree of success in drawing adherents. An early entry was the "popular health movement," led by Sylvester Graham (1794–1851), a temperance speaker who in the 1830s began to lecture against gluttony, improper dress, sexual permissiveness, and medicines, while arguing in favor of bathing, fresh air, exercise, and alterations to diet. Graham maintained that man was heading toward physical degeneration by not living according to the dictates of Nature's laws. Some of his arguments appeared most reasonable. Bathing in this period was not regularly practiced, the common diet was unbalanced, and clothing for women was unnecessarily restraining. Although the farmer was constantly exposed to sunlight, fresh air, and exercise, the commercial and leisure classes were not; indeed, many believed such a life to be unwholesome or demeaning. On the other hand, much of Graham's advice, notably his ramblings on the supposed evils of too frequent sexual encounters, was based upon what one biographer has called a "sublimated Puritanism."[41]

In 1839 he published a collection of lectures which became a best seller. Graham claimed that following the principles outlined in his book would make drugs and physicians unnecessary. First, adherents would be less likely to get sick; second, if they did fall ill, they would not be as severely affected; and third, by allowing their bodies' natural self-restorative powers to operate, they would recover more quickly.[42] Graham ruled out certain "unhealthy" foods: meats, fresh milk, eggs, coffee, tea, and pastries. His substitutes were invariably bland and tasteless; the best known of these was a cracker that still bears his name, originally designed to curb not only one's hunger but also one's sexual appetite. Graham's critics were quick to point out that his ultimate goal seemed to be to take enjoyment out of not only the kitchen but the bedroom as well.

Though Graham argued that his system was all inclusive, a number of his followers were among those who soon began to frequent the offices of another group of drugless practitioners, the hydropaths. An Austrian peasant by the name of Victor Priessnitz (1799–1851) had discovered that cold water seemed quite effective in treating many of the chronic ills of both man and beast, most notably gout and rheumatism. Within a short time his approach caused a small sensation in Europe, and several sanitaria were opened on the Continent for the teaching and practice of his methods.

Hydropathy was exported to America in the 1840s. Proponents founded two medical schools, and by the mid-1850s at least twenty-seven spas were in business, mostly in rural areas of the East and Midwest, where the water was believed to be the purest. The cure primarily consisted of drinking the precious fluid as well as enveloping one's body in it. According to Marshall Legan, "A sheet of cotton or linen dipped in cold water was spread on several thick woolen blankets. . . . Over the whole was thrown a feather bed, and the patient remained in his cocoon from twenty-five minutes to several hours, depending upon the seriousness of his condition and his ability to work up a good perspiration. Next the victim was unswathed and cold water was poured over him, or he was immersed in a cold bath and finally briskly rubbed dry."[43] Quite clearly, "heroic" therapy could be practiced by drugless healers as well.[44]

Still's familiarity with Graham's notions and hydropathy can be traced to his early manhood, when a utopian colony following a combination of these ideas was established near the Shawnee Mission.[45] This experimental community did not last long, but while it existed the Reverend Still had to be called on several occasions to care for those not responding to, or suffering from, the regimen. Undoubtedly Andrew was not especially impressed then or later with these methods. Yet Still came to believe that the drugless approach was the right one. It was only a matter of seeking out a system that could provide a more logical basis for reliable diagnosis and efficacious treatment. Towards this end, Still would find considerable guidance in the principles and practices of magnetic healing.

In 1774 Franz Mesmer (1734–1815), an Austrian physician, postulated that an invisible universal magnetic fluid flowed throughout the body and that too much or not enough in either a part or the whole was one major cause of disease, particularly nervous disorders. The only rational course of treatment, therefore, was to restore the fluid to its proper balance. This could be accomplished by making passes over the body with magnets or his hands. Mesmer was not the first to heal through the use of touching; rather he was the first to fashion this approach into a coherent system of medical practice.[46]

Many of his early cures through this method were greatly publicized, and soon he was attracting patients from all across Austria. His success there was short-lived, however, as pressure from the medical community of his native Vienna forced his departure for Paris. In the French capital, Mesmer's practices became more irregular. Instead of seeing clients separately and discreetly, he formed groups and ministered to several patients at once. Often he employed a huge indoor tub with extended "magnetized" rods. Those gathered would bathe together, placing the afflicted parts of their bodies against the metal protrusions, until Mesmer materialized. While an orchestra played solemn music, he entered the room dressed in a flowing, lilac-colored robe and touched his patients as he passed. This was designed to bring each individual to a near seizurelike state, which, according to Mesmer, was often necessary to achieve catharsis. The tub was not his only healing site. Clients would also be treated outdoors, under "magnetized" trees or beside "magnetized" rocks.[47]

In 1784, as Mesmer's practice gained popularity, two special commissions were created to investigate the relative merits of his claims. One of these groups was appointed by the French Academy of Sciences and included in its ranks Benjamin Franklin, Jean Sylvan Bailly, and Antoine Lavoisier. This committee declared that a "magnetic fluid" did not exist, that Mesmer's cures were only the result of suggestion, and that the morals of women undergoing such treatment were being threatened. In an induced seizure, they argued, females could become easy targets for seduction.[48] With the appearance of this study, Mesmer's personal influence waned. Some of his followers, believing his basic principles to be valid, abandoned the tub and other questionable procedures and continued the attempt to gain respectability. In succeeding decades, they made progress. In 1831 a somewhat favorable report on the subject was issued by the French Academy of Medicine, and backhanded support came later, in the writings of James Braid (1795–1860) on what was eventually called hypnosis.[49]

Magnetic healing was brought to the United States in 1836 by Charles Poyen (d. 1844), who gave a series of public lectures in Boston and took on a number of students, training them in massage and other methods then thought to be useful in restoring fluid balance.[50] Poyen's activities helped stimulate considerable interest in magnetic healing, and though his stay in America was relatively brief, the seed he planted was soon able to sprout without him.[51] One of those who allegedly heard Poyen lecture was Phineas Parkhurst Quimby (1802–65), who afterwards established a practice consisting largely of verbal suggestions combined with light stroking of the body. Though Quimby's writings were not published until after his death, he was an influential figure during his lifetime, serving as physician, teacher, and inspiration to Mary Baker Eddy (1821–1910), the founder of Christian Science. Quimby's ideas also constituted the intellectual fountainhead for the loose confederation of religious groups known as "New Thought."[52]

The best-known magnetic healer prior to the Civil War was Andrew Jackson Davis (1826–1910), who was also the leading American exponent of spiritualism.[53] In the first volume of his massive tome, *The Great Har-*

monia (1850), Davis sought to combine both belief systems. Conceiving of the body as a machine, he maintained that health was simply the harmonious interaction of all man's parts in carrying out their respective functions. This was due to the free and unobstructed flow of "spirit." Any diminution or imbalance of this "fluid" would cause disease. Like others before him, Davis placed emphasis on healing with his hands. Of particular interest is his management of asthma, which consisted in part of vigorous rubbing along the spinal column.[54] While this type of treatment constituted but a small feature of Davis's practice, later magnetic healers, perhaps influenced by the attention given the spine by such orthodox European physician researchers as Charles Bell (1774–1842), François Magendie (1783–1855), and Marshall Hall (1790–1857), made extensive use of manipulation. One of these was Warren Felt Evans (1817–89), whose name is most often associated with "Mind Cure."[55] In his book entitled *Mental Medicine* (1872), which went through fifteen printings, Evans noted, "By the friction of the hand along the spinal column, an invigorating, life-giving influence is imparted to all the organs within the cavity of the trunk. The hand of kindness, of purity, of sympathy, applied here by friction combined with gentle pressure, is a singularly effective remedy for the morbid condition of the internal organs. It is a medicine that is always pleasant to take."[56]

These sentiments were echoed in the book *Vital Magnetism* (1874), written by another popular healer, Edwin Dwight Babbitt (1828–1905). He specifically mentioned convulsions, apoplexy, sunstroke, headache, muscular complaints, common rheumatism, and paralysis as disorders capable of cure through spinal treatments.[57] It is not known whether Still read these works by Davis, Evans, and Babbitt; however, he was well aware of their message. A letter cosigned by him to the editors of the *Banner of Light* indicates that he was a reader of that journal, which was oriented toward spiritualism and magnetic healing and published articles and advertisements by those practitioners within its pages.[58]

Though Still never embraced all of the ideas of such contemporaries, a number of the central tenets of magnetic healing made a strong impression on him: the metaphor of a man as a divinely ordained machine; health as the harmonious interaction of all the body's parts and the unobstructed flow of fluid; and, of course, the use of spinal manipulation. His most significant departure from them would be over the nature of the fluid. While for the remainder of his life he spoke obliquely of the physiological role of magnetic energy, it was free flow of blood, he believed, that constituted the key to health.[59] "I proclaimed," he later wrote, "that a disturbed artery marked the beginning to an hour and a minute when disease began to sow its seeds of destruction in the human body. That in no case could this be done without a broken or suspended current of arterial blood itself. He who wished to successfully solve the problem of disease or deformity of any kind in every case without exception would find one or more obstructions in some artery or vein."[60]

In June of 1874, Still severed his ties to regular medicine, an action that shocked his community. Many of his friends and relatives, in response to his odd theories and particularly his "laying on of hands," questioned his sanity. The local minister, seeing him as an agent of the devil, had him "read out" of the Methodist Church. Still asked for permission to explain his practice at nearby Baker University, a school he had helped establish, but the privilege was denied.[61] Effectively ostracized in Baldwin, Still traveled to Macon, Missouri, to visit a brother and see if public acceptance of his newly adopted ideas and methods would be any better there. It was not. After staying a few months, treating but a small number of patients, he moved on to Kirksville, situated in northeast Missouri, where, to his surprise, "three or four thinking people" actually welcomed him.[62] The city then had a population of eighteen hundred and was the commercial capital of Adair County, which had a total of some thirteen thousand inhabitants. In a local paper, the *North Missouri Register*, he advertised himself: "A. T. STILL, MAGNETIC HEALER, Rooms in Reid's building, South Side Square, over Chinn's store. Office hours—Wednesday's, Thursday's, Friday's, and Saturday's from 9 a.m. to 5 p.m. with an intermission of one hour from 12 p.m. to 1 p.m."[63] Though his practice in this new locale was not successful at first, he was comforted that there was no organized harassment by either the clergy or the local physicians. Still was also able to go about his business without serious interference from the state. In August 1874, while in Macon, he registered with the county clerk as a physician and surgeon, thereby protecting himself from prosecution for practicing without a license.[64] As a result of the initial tolerance he had been shown in Kirksville, he moved his family there the following May.

In the fall of 1876 Still contracted typhoid fever, the effects of which made him an invalid for more than six months. After he fully recovered, Still realized that his local clientele would be too small to support his loved ones as well as pay off the debts incurred during his illness. In desperation he applied for a federal army pension but was turned down, since his service in the Civil War had been entirely with state militias.[65] With few Kirksville patients, it was necessary to expand his population base. Still became an itinerant practitioner. He traveled for extended periods to several communities throughout the state, while his wife and children remained in Kirksville. For the next few years Still's earnings barely kept pace with his expenses. On various occasions his relatives offered to help him out financially if he would return to orthodox medicine, but he adamantly refused.[66]

THE LIGHTNING BONESETTER

Sometime during the 1870s Still became interested in bonesetting, another form of manipulative practice generally limited to the field of orthopedics. In deciding to learn these techniques, he may have hoped to be able to treat a wider range of disorders, thus giving him the potential of substantially increasing his patient load and his income.

Bonesetters were an ancient if not respectable group of healers. In England they had enjoyed a relatively unfettered practice among the common people, who could not afford a regular physician and who often had difficulty locating one willing to treat them. However, bonesetters could also count on the patronage of members of the upper classes, including royalty. It was widely believed that their particular talent was passed down from one family member to another and was therefore a gift that transcended formal book learning.[67]

In addition to reducing dislocations, bonesetters also manipulated painful and diseased joints, thinking that these conditions were also caused by a "bone out of place." Physicians ridiculed their crude diagnoses and dismissed their claim that such treatment was of any value. Nevertheless, some patients with restricted joint mobility that remained unrelieved after treatment by trained orthopedists apparently benefited from manipulative therapy administered by such "quacks." Some physicians assumed that these clients were only hysterics, or that the patient and the bonesetter were in collusion to embarrass the doctor in charge; but in 1867 Sir James Paget, himself a most distinguished surgeon, startled his colleagues by announcing that he believed there were joint maladies that bonesetters, regardless of the inaccurate diagnoses, were able to cure and that only through a searching investigation of their techniques could the relative value of such treatment be fully understood. "Few of you," he admonished his educated brethren, "are likely to practice without having a bonesetter for a rival and if he can cure a case which you have failed to cure, his fortune may be made and yours marred."[68]

In 1871 Dr. Wharton Hood, an acquaintance of Paget, published a book in England and the United States based upon his experiences as a bonesetter's apprentice. As he described it, the bonesetter's technique constituted "the art of overcoming by sudden flexion or extension, any impediments to the free motion of joints that may be left behind after the subsidence of the early symptoms of disease or injury." The conditions for which Hood believed this type of treatment useful were: cases of stiffness, pain, and adhesion following fractures and sprains of one or more of the bones forming a joint; rheumatic or gouty joints; displaced cartilage; subluxations of the bones of the carpus and tarsus; displaced tendons; hysterical joints; and ganglionic swellings.[69] However, he cautioned that bonesetting was only successful where the ability of joints to rotate had not already been permanently destroyed.

Hood noted that, while most bonesetters' activities were limited to manipulating the extremities, they were also treating people who were "complaining of a 'crick' or pain, or weakness in the back, usually consequent upon some injury or undue exertion, and . . . these applicants are cured by movements of flexion and extension, coupled with pressure upon any painful spot." Often during these maneuvers a "popping" or "clicking" sound would be emitted by the spinal joints, which many times convinced the patient that a cure of the problem had been effected.[70]

Bonesetters had been found in America since the colonial era; the most prominent practitioners were the Sweet family, who held forth in the New England area for nearly two hundred years.[71] How widespread such manipulators were elsewhere in the country is unclear. One physician in 1884 estimated that in every city in the United States there were "individuals claiming mysterious and magical powers of curing disease, setting bones, and relieving pain by the immediate application of their hands."[72]

It is not known how Still learned to become the "lightning bonesetter" he would throughout the 1880s advertise himself to be. Though he could have come across Hood's book, it seems more likely that his knowledge was derived from observing the work of another practitioner in the field.[73] However he learned these methods, Still soon afterwards made an important discovery, namely, that the sudden flexion and extension procedures peculiar to this art were not limited to orthopedic problems and that they constituted a more reliable means of healing than simply rubbing the spine. He would later recount this story from about 1880: "An Irish lady . . . had asthma in bad form, though she had only come to be treated for the pain in her shoulder. I found she had a section of upper vertebrae out of line, and I stopped the pain by adjusting the spine and a few ribs. In about a month, she came back to see me without any pain or trace of asthma. . . . This was my first case of asthma treated in the new way and it started me on a new train of thought."[74] Soon he was handling headache, heart disease, facial and arm paralysis, lumbago, sciatica, rheumatism, varicose veins, and an increasing variety of other chronic ailments, all by manipulating vertebrae back into their "proper position." In accounting for his success, Still would synthesize some of the major components of magnetic healing and bonesetting into one unified doctrine. The effects of disease, as the former said, were due to the obstruction or imbalance of the fluids, but this in turn was caused by misplaced bones, particularly of the spinal column which interfered with nerve supply regulating blood flow. Still had given birth to his own distinctive system.

In the next decade Still traveled across Missouri touting his new approach. According to one eyewitness:

> Sometimes he would leave Kirksville with barely enough money to pay carfare and go to some town with a bundle of probably a thousand [hand]bills, get them scattered, after which he would give an exhibition of setting hips, probably on a public square, in a spring wagon or ox-cart. Of course, he would be looked upon as some mysterious being, crazy, or at least daffy; but with his intuitive insight, he would pick out a cripple, or someone with a severe headache or some disease that would cure quickly, and demonstrate before the anxious crowds.[75]

Often, Still had a difficult time getting his ideas across to potential clients. He saturated his speeches with an odd collection of metaphors, parables, and allegories which left many listeners bewildered.[76] His unusual attire—a rumpled suit, a slouch hat, his pants tucked inelegantly in his boots—caused some to look rather than listen. Many times he could be seen on the streets clenching a long wooden staff he used as a walking stick while toting a sackful of bones over his shoulder. Not surprisingly, such behavior led people to decide that he was either an eccentric genius or a deranged old man. Nevertheless, as one follower noted, "the impression left was usually a good one."

One or more of his sons would often travel with him and assist in treatment. Harry Still later observed:

I believe I would be safe in saying that the six months we practiced in Hannibal [winter 1884–1885] we accumulated a dray load of plaster of paris casts, crutches, and all classes of surgical appliances [no longer needed by patients]. We went from Hannibal to Nevada, Missouri where the state insane asylum is located. Here we made fully a hundred cures. . . . I remember one interesting case. The lady had been in the asylum for several years. It seemed she had lost her mind suddenly while playing a piano. Father examined her neck and found a lesion of the atlas. In less time than I have taken in telling, the girl was as rational as ever. Strange to say, the first thing she said was "Where is my piano and music?" She was anxious to finish the piece she had started playing three years before."[77]

Such unusual recoveries gradually spread the "lightning bonesetter's" reputation. By the late 1880s his scheduled trips to various towns would cause considerable local excitement. In Eldorado Springs he had to reserve sixteen rooms to treat the crowds that had gathered. People reportedly came to Nevada City from upwards of 150 miles away, complete with tents in anticipation of a long wait.[78] Still had become a charismatic figure.

Paradoxically, it was only after he had obtained notoriety elsewhere that the people of Kirksville began to patronize him in large numbers. One incident in particular helped change his image. The young daughter of the town's Presbyterian minister, J. K. Mitchell, had for some unreported reason lost her ability to walk. Mrs. Mitchell, after her child had been treated without apparent benefit by other local physicians, in desperation asked her husband to allow Still to make an examination. The clergyman adamantly refused. Nevertheless, while her husband was away on an extended trip, the child's mother called for Still, who proceeded to adjust the girl's spine. When Mitchell returned home his daughter walked down the stairs to greet him. The good pastor thereafter sang praises to Dr. Still's name, and the social barriers that had long prevented "the lightning bonesetter" from treating "decent folk" were lowered.[79]

As a result of his newfound respectability, Still decided to make Kirksville his permanent base of operations. In 1889 he established an infirmary. Soon patients from great distances were seeking him out. "It was a problem," said one follower, "how best to take care of the people that were flocking to him. . . . Many of those who came were pronounced hopeless by other physicians. Some of them were hopeless. But he was able to cure enough . . . to keep adding to his reputation and his fame which extended into ever widening circles."[80]

His success convinced Still that he had discovered a new science of healing. He lacked only a proper designation. "I began to think over names such as allopathy, hydropathy [and] homeopathy," he recalled. Eventually this led to "start out with the word *os* (bone) and the word pathology, and press them into one word—*osteopathy*."[81]

THE MISSOURI MECCA

Having named his new system of medicine, Still decided it was time to share his discovery with others. In 1892 he opened the American School of Osteopathy, charging his students $500 for several months of personal instruction. Upon completion of the specified course, students were awarded a certificate stating they were "diplomats in osteopathy," or DOs. Within six years, the school changed the title on the certificate and bestowed on new graduates the degree Doctor of Osteopathy.

A few months before his first classes were to begin, Still had the good fortune to meet Dr. William Smith (1862–1912), a thirty-year-old Scottish physician who was in town on a business trip. Smith had been trained at Edinburgh and had studied for several additional years on the Continent; his background was in stark contrast to that of the self-taught country doctor.[1] After hearing Still spoken of as the "d——d quack" by a local regular physician, he decided to investigate osteopathy on his own. "I sat entranced," Smith wrote a few years later of his first encounter with Still. "The theories he introduced were so novel, so contrary to all I had read or heard that I failed to follow his reasoning. Arguments as to their impossibility were simply met with one statement: 'But it is so; there are no "ifs" and "ands" about it, I do what I tell you and the people get well.'"[2] After visiting several boarding houses around town and seeing the results that had been obtained under such care, Smith became convinced that something of value was being imparted. As he wanted to learn more, Smith accepted Still's offer to teach him everything he knew; in return, the young doctor would serve as an instructor in Still's proposed school.

On October 3, 1892, classes began, with upwards of twenty-one students, men and women, enrolling at the beginning or during the first several weeks of the course. The ages of the students ranged from eighteen to sixty-five. Some held college degrees, others had nothing more than a common school education. All of them, however, had been direct or indirect beneficiaries of Still's ministrations. Each morning for four months Smith drilled the group in anatomy without the benefit of a cadaver upon which to demonstrate. This problem was compensated for in part by his lecturing, which, according to one of his students, "was of such an impressionable type that one who listened to him could virtually look into the human body with his mind's eye and see all its numerous functions."[3] Smith's role was also symbolic. As one early student shrewdly noted, "'Bill' furnished the 'front.' He 'looked good' to the people and inspired confidence in infant osteopathy."[4]

Following their daily anatomy lessons, students spent the afternoons in the infirmary with the "old doctor," as he was affectionately called. Still was a natural philosopher rather than an academician. Students had to pick up what knowledge they could by listening to his extended metaphors and his sometimes rambling commentary. One of his followers declared:

> He rose to the lofty heights of his conceptions of life, health, disease and medicine by the purest of intuition. He wiped the slate of knowledge, as it were, of much if not most of the accepted, accredited teachings of the day, not only in the field of medicine, but also in science, religion, ethics, politics, and endeavored to begin his thinking upon any and every subject with the new data of pure forms, built out of his imagination, with little regard or discomfort if his excursions took him sheer in the face of every accepted belief and profession.[5]

"The human body," Still told his students, "is a machine run by the unseen force called life, and that it may be run harmoniously it is necessary that there be liberty of blood, nerves, and arteries from their generating point to their destination." He then illustrated the significance of this basic principle with a colorful analogy:

> Suppose in far distant California there is a colony of people depending upon your coming in person with a load of produce to keep them from starving. You load your car with everything necessary to sustain life and start off in the right direction. So far so good. But in case you are side-tracked some where, and so long delayed in reaching the desired point that your stock of provision is spoiled. If complete starvation is not the result, your friends will be at least poorly nourished. So if the supply channels of the body be obstructed, and the

First class of the American School of Osteopathy. In center, holding cane, is Dr. A. T. Still; on other side of skeleton is Dr. William Smith. *Courtesy of Still National Osteopathic Museum, Kirksville, Missouri.*

life-giving currents do not reach their destination full freighted with health corpuscles, then disease sets in.[6]

Given such circumstances the osteopath would "remove the obstruction by the application of the unerring laws of science, and the ability of the artery for doing the necessary work will follow. As a horse needs strength instead of a spur to enable him to carry a heavy load, so a man needs freedom in all parts of his machinery with the power that comes from the perfection in his body, in order to accomplish the highest work of which he is capable."[7]

The highlight of the students' day was watching him operate. According to one, "We would hold the patients in position while Dr. Still . . . worked upon them, explaining to us as he treated why he gave this movement in one place, and a different movement in another. He would tell us what it would mean to the nerves from that particular region if muscles were 'tied up' or a bone was out of line." In diagnosing these conditions, a student explained, Still taught "that we should place the patient on his side and then pass our hands carefully over the spinal column from the base of the spine, noting temperature changes as we went along. Should

THE DOS: OSTEOPATHIC MEDICINE IN AMERICA

there be a lesion along the spine, where nerves may be disturbed, it would easily be detected through an abnormal coldness or hotness of the tissue at that point."[8]

Students often had the opportunity to observe the progress of patients over an extended period of time, because the afflicted usually agreed to remain in Kirksville for a minimum of one month. "Remember," Still told his clients,

> when many of you come to me you are not the most choice kind of patients. Remember the company you have kept before coming here. You have been with doctors who blister you, puke you, physic your toenails loose, fill your sides and limbs with truck from hypodermic syringes. You come to me with eyes big from belladonna, back and limbs stiff from plaster casts—you have been treated and dismissed as incurable by all kinds of doctors before coming to us, and if we help you at all—we do more than others have done.[9]

While in Kirksville each supplicant had to obey the interdiction against the use of liquor. "We do not wish to treat habitual whiskey tubs," Still declared. "This rule must be strictly obeyed by all patients, and those who feel they cannot conform to it had better stay away." Internal medications were taboo as well: "No system of allopathy, with its fatal drugs should e'er be permitted to enter our doors. No homeopathic practice with its sugar coated pills, must be allowed to stain or pollute our spotless name. . . . Osteopathy asks not the aid of anything else. It can 'paddle its own canoe' and perform its works within itself when understood. All it asks is a thorough knowledge of the unerring laws that govern its practice and the rest is yours." Still left no doubt that he considered alcohol and other drugs a moral evil. Indeed, often his crusade centered more on them than on the benefits of manipulation: "Was God ever drunk? Was Nature ever intoxicated?" He once estimated that 90 percent of his work involved overcoming the effects of such poisons upon the body.[10] If one had faith in the wisdom and completeness of God's design, Still maintained, one had to see that the use of drugs was not just immoral, it was unnecessary as well.

THE WORD SPREADS

Throughout Missouri and elsewhere, more people were beginning to learn of Still's activities from sensationalized newspaper accounts. The *St.*

Louis Democrat called Kirksville "the great Mecca for invalids, particularly those suffering from bone disease, dislocations, and similar afflictions. To and fro there surges a throng of ailing humanity sincere in purpose as the Eastern devotee who kneels at the tomb of Mohammed. But the results accomplished are not visionary or fanciful, they are real and practical. Marvelous even unto the miraculous are some of the cures and yet they are all treated in a natural and scientific manner." The *Des Moines Daily News* noted in its columns that Still was "performing remarkable cures in a very simple way," while the *Nebraska Daily Call* declared that osteopathy had deservedly "won a substantial claim to the confidence of all classes of invalids." Some reporters told their readers that they had arrived in Kirksville hardened skeptics but left true believers. One Iowa journalist confessed, "It was an experience for the correspondent which removed from his mind every vestige of incredulity to the truth of these innumerable testimonials favorable to osteopathy and the eminent doctor." The editor of the *Bethany (Illinois) Echo* even submitted himself for treatment and later told his readers, "If you have an ailment which our doctors cannot successfully treat we advise you to go to Kirksville and be cured.[11]

All of these and other favorable stories were subsequently reprinted in a monthly tabloid Still published called the *Journal of Osteopathy*, which was mailed to friends and relatives of patients in Kirksville and to local papers throughout the Midwest. Average issue circulation rose from several hundred in 1894 to more than 18,000 two-years later.[12] With this publicity greatly increasing the number of patients, the railroads scheduled more trains through the town to accommodate the traffic and established special reduced-rate fares for those who required shuttle service. Entrepreneurs built new hotels, and boarding houses flourished. Storekeepers were pleased to find that their shops were constantly crowded. Seemingly everyone in town was prospering from Still's work.

Still's representatives met each train that pulled into the station, greeting clients, arranging accommodations, and setting up appointments with his staff. One reporter noted of the infirmary:

Everything is managed as smooth as clockwork. . . . The almost constant ringing of electric bells announcing that room so and so is ready for another patient, the great discipline with which patients take their positions near the doors of the operating rooms when their turn is "next," the incessant click of the typewriter as it wades through the immense correspondence, the frequent "helloing" at the telephone, and the general counting room appearance of the business office, impress the visitor that besides an understanding of the human mechanism and laws of health, a thoroughly organized business system is required to do the great amount of work accomplished daily by this wonderful institution. There are now over five hundred patients and when it is remembered that treatment is given each patient from one to three times a week, it is not difficult to understand that the ten operators are kept moving.[13]

Of those who came to the infirmary, one journalist observed:

> Almost every phase of society, nearly every section of the country, and certainly quite, if not all the ills to which human flesh is heir, were represented. There was the laboring man, the business man, and the professional man; there was the working girl, and the society favorite; there was the anxious husband with the invalid wife, the loving mother with her crippled child; there were scores on crutches and in invalid chairs; there were others who were compelled to depend on strong arms and tender hands. One thing they possessed in common, and that was a beaming countenance that indicated confidence, an expectancy, if not already a realization of a bettered condition.[14]

Contemporary accounts indicate that the majority of incoming patients were suffering from chronic, noninfectious disorders. In a sample of forty-nine patients cited in one issue of the *Journal of Osteopathy*, ten were diagnosed as having some form of joint dysfunction, seven a nervous disorder, six asthma, five partial or complete loss of sight, hearing, or voice, three bowel difficulties, and the remainder had other long-term health problems.[15] One enterprising patient took a survey of 109 fellow sufferers. Of these, 61 percent reported having some form of "spinal complaint"; the second most common problem was bone or joint maladies of the extremities. A different point of view was provided by a reporter from *Godey's Magazine*, who wrote, "From my own observation I think that a majority of the patients were afflicted with nervous trouble and paralysis."[16]

Many clients were willing to discuss their cases openly and give testimonials to the newspapers. F. H. Barker was a Methodist minister from Kansas who had fallen from a train, severely wrenching his neck; he subsequently developed sore eyes and ultimately became blind. Barker claimed to have seen oculists in three states before coming to Kirksville. After undergoing treatment at the infirmary for five weeks he reported that he had no more pain and that his eyesight was almost normal. R. W. Neeley of Franklin, Kentucky, developed a serious case of "heart disease with nervous prostration." "When I landed in Kirksville," he told the press, "I could not walk across the room without holding to chairs. I felt like toppling over at every step. From the very first treatment I began improving, and can't express it better than to say I feel like a young colt in a clover field on a bright spring morning." Mrs. J. T. Christian, the wife of a Baptist preacher, brought her seven-year-old son, who had suffered for three years from what had been diagnosed by two physicians as hip joint disease. According to the newspaper account, "As soon as he was placed on the operating table here a partial dislocation of the hip and spine were discovered. They were at once reduced without weights, braces, plaster of Paris, or any other paraphernalia; and now the boy is able to go anywhere on crutches, without his brace, feels no pain, the abscesses having disappeared, and will soon be well again." After being treated for several weeks for a chronic case of sciatica, J. W. Blocker of Dark County, Ohio, was only bothered by an occasional pain in his ankle. Mr. English from Quincy, Illinois, had been deprived of the full use of his right leg and arm and other parts of the body by what was diagnosed as a "spinal affection." After treatment he was walking without crutch or cane. Asked by a reporter if he believed in miracles, Mr. English replied, "Not often, but I am a firm believer in osteopathy."[17] Perhaps the most publicized early case was that of the son of U.S. senator Joseph Foraker (R-Ohio), who was sent to Kirksville with what had already been diagnosed as "valvular disease of the heart." As his physicians had given up any hope for the child's survival, there seemed little to lose by trying Still; and indeed, under his care, the symptoms of the condition gradually disappeared. Because of the national attention given the Foraker child, several among the political and economic elite came to Kirksville and then further spread osteopathy's good name.[18]

IN THE LEGISLATURE

The events taking place in Kirksville did not escape the attention of the Missouri State Medical Association, which determined to put a stop to them. The association's first action appears to have been taken as early as 1889, when it pushed through the legislature an amendment to the existing healing arts law which read, "Any person who shall by writing or printing or any other method publicly profess to cure or treat diseases, injuries, or deformities by manipulation or other expedient, shall pay to the state a license of $100 a month." This, however, was unenforced.[19] Later, in

1893, after Still had secured his school charter and begun teaching his first class, the regular physicians, in cooperation with the homeopathic and eclectic medical societies, introduced a bill into the House rendering it necessary for those practicing osteopathy to be graduates of a reputable medical school. Apprised in advance of this move, Still's students and patients began a petition drive that was active throughout the state. Hundreds of protesting letters and telegrams poured into the lawmakers' offices, leading to the defeat of the bill by a wide margin.[20]

After winning this battle Still and his followers went on the offensive, seeking specific legislation that would guarantee the legal right of DOs to practice within Missouri's borders. One of Still's legal advisors, P. F. Greenwood, linked osteopathy to the aspirations of Midwestern populism. In reference to the three established schools of medicine, Greenwood drew a religious analogy: "Suppose Baptists, Methodists, and Cumberland Presbyterians were the only recognized churches to save souls in this state and we were assured the legislature intended to rid the people of the Commonwealth from the doctrines and teachings of heretics? Would you call that class legislation? A monopoly of free gospel certainly. Then is not our medical class legislation as bad? I hold that if medicine is a science that no legislation is necessary to uphold or protect it." Greenwood continued his defense of osteopathy by stating, "it is a science, and all it asks is an equal chance in the race of life. If it is not a science the challenge is open to the world to disprove it. It asks but one favor and that is the modification and change of the unfair medical laws of this state."[21]

Leading the medical opposition was a prominent orthopedic specialist from St. Louis by the improbably coincidental name of A. J. Steele. He argued that every cult, regardless of its methods, professes cures. Osteopathy was simply no different from Christian Science, magnetic healing, and the water of Lourdes. As to osteopathic theory, it was entirely invalid. Steele pointedly asked:

Is the honest, scientific work of educated men and acute observers of the past ages down to the present to be thus ruthlessly set aside? Do our studied research in pathology and therapeutics go for naught? Strange is it not that of the thousands of skeletons carefully examined that frequent examples of misplaced bones have not been discovered, if such truly is the cause of all disease? We see patients daily recovering from sickness and disease in whom no effort has been made to reduce misplaced bones, showing that the *causus morbi* did not lie in that direction. *Per contra*, we have had cases where dislocation and deformity did exist, for example of the spine, and neither organic nor special disease followed—the soft parts accommodating themselves to the displaced bones and the normal functions being well performed.[22]

Some osteopathic followers responded by challenging this critic's competence, claiming that a number of his unsuccessfully treated patients had come to Kirksville only to return home free of their affliction.[23] William Smith, however, admitted that there was due cause for the skepticism of his fellow MDs.

If a man, a physician, comes to Kirksville and hears what he will hear and tries to reason it out on the basis of what he learned in medical school, there is only one conclusion to which he can come: that osteopathy is a fraud and a delusion, a gigantic humbug which is taking from the pockets of the sick and afflicted thousands of dollars monthly. BUT, if the enquirer will just approach the matter as though he knew nothing (and after four years experience of osteopathy let me tell any doctor that he knows very little), take nothing for granted, accept no statement for or against osteopathy; but just interview a dozen patients and accept them as reasonable men and women and not as hysterical persons, half-fitted for the lunatic asylum, nor utter and gratuitous liars, he is BOUND as an honest man to come to the conclusion as I did that there are still some things in the healing art which are not known to the medical profession. Let him examine further and he will find results obtained quite impossible under treatment with medicine. Then let him inquire of the patients who tell him in their stories, how many doctors had declared their recovery impossible, and then, and not until then, let him make up his mind as to whether or not osteopathy is a fraud, its practitioners humbugs and its supporters liars. If all these persons claiming to be benefited are liars where can the profit come from in running the business? To pay such an army of liars would consume the capital of the state. If they are hysterical why did not their doctors cure them?[24]

Despite vigorous medical opposition, the legislature ultimately voted in favor of the osteopathy bill. This measure called for DO graduates to present their diplomas for registration to their county clerk, who in turn would issue a certificate making them eligible to treat disease through the hands. The osteopath would not be required to pass any test or attend classes for any set length of time.

All of this went for naught. When the proposed law reached the desk of Governor William Stone he vetoed it on the grounds that osteopathic practitioners were insufficiently educated. "Medicine is a science," he declared. "A judicious practice of it requires a good general and fundamental education, and a thorough knowledge of all the departments of medicine: anatomy, physiology, chemistry, pathology, therapeutics, practice, etc."[25] In other words, if osteopathy wanted equal treatment under the law, it had to conform to the academic scope adhered to by other practitioners of the healing art.

The bill's advocates excoriated Stone, but the substance of his objections went unanswered. Indeed, there was little Still or his followers could say to justify what then constituted osteopathic education and standards. From 1892 through 1896, three classes had graduated. The length of training varied from nine to eighteen months and consisted of lectures in anatomy, osteopathic principles and technique. Still believed other subjects were unnecessary. Once, he stormed into class, raced to the blackboard, and wrote on it in large letters "NO PHYSIOLOGY" and then left the room. Anatomy, in his view, was the sole medical certainty. There was no need to bother with the theories and speculations of other branches of medicine.[26] The first charter of the American School of Osteopathy and early board meetings show that Still wanted to include surgery as well as obstetrics as part of the curriculum, but at that time these subjects had not been introduced.[27]

Stone's veto and the urgings of Smith and others who had argued that the DOs training was incomplete finally convinced Still that he had to make changes. By the end of 1896 he had formally lengthened the course of study to four terms of five months each, and at dedication ceremonies of a new college building he announced, "I am now prepared to teach anatomy, physiology, surgery, theory and practice, also midwifery in that form that has proven itself to be an honor to the profession."[28] Several months later the school published a more detailed course outline that also included histology, chemistry, urinalysis, toxicology, pathology, and symptomatology.[29] Thereafter Still's supporters maintained that every subject covered in a standard medical college, with the exception of *materia medica*, was taught at the American School of Osteopathy.

After Still had thus complied with the governor's major objections, his followers revised their bill and resubmitted it to the legislature, which speedily passed it. This time, the DOs did not face an executive veto. In the interim, Stone had left office; and his successor, Lon Stephens, an osteopathic patient, gladly signed the measure into law on March 3, 1897.[30] When word reached Kirksville, pandemonium reigned while the entire population set aside a day for celebration. "The morning was ushered in," according to one newspaper account, "by the firing of anvils in honor of Governor Stephens, the legislature, Dr. Still and everybody connected with the fight. Bells rang and whistles blew. Anything that would make a big noise went. Residences, stores, shops were decorated, the big osteopathy building was covered with flags and bunting inside and out, and the whole city donned its best fourth of July attire."[31] As students marched down the streets they cheered:

> *Rah! Rah! Rah!*
> *Missouri passed the bill*
> *for AT Still*
> *Good-bye Pill*
> *We are the people*
> *of Kirksville.*[32]

THE NEW FACULTY

The passage of this law brought more matriculants, now confident that their time and money would not be wasted. Within a few years there were seven hundred full-time students in attendance.[33] This growth, coupled with the expansion in the curriculum, forced Still in 1897 and 1898 to engage additional teachers. Among those he found to assist Smith and himself were C. W. Proctor (1859–1949), holder of a PhD in chemistry; Charles Hazzard (1871–1938), a university graduate who also held a DO; Carl McConnell (1874–1939), who after earning his osteopathic diploma received an MD at a homeopathic school; and the three Littlejohn brothers: J. Martin (1865–1947), president of Amity College in Iowa, who had law and divinity degrees from the University of Glasgow and a PhD in political science from Columbia; James (1869–1947), who held both an MD and a master's degree in chemistry from Glasgow; and David (1876–1955), who earned an MD degree from a Michigan medical school. All of these new faculty either had benefited personally from osteopathy or had a close relative or friend who had.

Although this group remained intact for a comparatively brief period, its effect upon the development and course of the movement was significant. Whereas Still had built his system largely upon the principles and practices of magnetic healing and bonesetting, his new faculty relied upon more reputable sources of knowledge.[34] Joint manipulation had a lengthy

American School of Osteopathy Infirmary Building with annexes, completed 1897. *Courtesy of Still National Osteopathic Museum, Kirksville, Missouri.*

orthodox tradition, and others before Still had postulated that disturbances or displacements of vertebrae could cause symptoms elsewhere in the body.

In ancient Greece, frictions—a form of massage—were employed to treat a wide range of ailments. Some of the Hippocratic writings deal extensively with the subject, one work noting that "the physician must understand many things and frictions not the least of all . . . for frictions can bind a joint that is too loose, and loosen a joint that is too rigid." In later centuries manipulation was practiced by Roman healers, but when the Empire declined, the art all but disappeared in Europe. Although rediscovered during the Renaissance, it took a minor position in therapy compared to drugging. However, over the course of the next several centuries it would be promoted by a host of distinguished physicians including Gerald von Swieten (1710–72), who advocated manipulation as a general measure to increase blood circulation. "The vital powers," he said in words that foreshadowed Still's, "may be increased by friction to any extent without any foreign addition to the body."[35]

Throughout the 1800s a small group of English and American doctors who employed massage in their practice tried to alert the medical profession to the modality's value through a number of books and articles. William Balfour (in 1819), for example, recommended it in rheumatism and sprains; John Bacot (1822) considered it helpful in treating several surgical diseases; and William Cleoburey (1825) manipulated in cases of contracted joints and lameness from various causes. Later in the century, S. Weir Mitchell (1872), as a result of his Civil War hospital experience, relied on manipulation to treat many traumatic nerve and muscle injuries and by 1877 was including it when treating neurasthenia, hysteria, and locomotor ataxia.[36] William Murrell (1886) added constipation, poisoning, lumbago, and sciatica to the list; G. L. Pardington (1886) migraine; and A. J. Eccles (1887), constipation.[37] The most extensive clinical research on the subject was carried out by Douglas Graham (1884), who, citing the results of fourteen hundred of his own cases as well as those handled by others, reported success in uterine disorders, hemiplegia, infantile paralysis, writer's cramp, muscular rheumatism, sprains and joint afflictions, rheumatoid arthritis, glaucoma, and in catarrhal affections of the nose, pharynx, and larynx.[38]

Some advocates of massage worked on spinal complaints. George H. Taylor (1884) noted:

> It has been clearly proved that the circulation of the blood, and therefore the proper nutrition of the spinal bones, are quite dependent on the flexibility of the spine, which displaces and replaces the vertebral nutritive fluids, much as the functional use of muscles secures their nutritive support. It follows that the proper therapeutics in vertebral disease is not suspension of the flexibility of the vertebrae by mechanical restraints. . . . It has been practically demonstrated that exactly the opposite course is therapeutically indicated and that the most successful treatment of vertebral disease consists essentially in judicious use of this physical property of elasticity and flexibility.[39]

While their practices paralleled those used by Still, neither Taylor nor other proponents of massage gave the spine any central theoretical role in disease; nor did they focus upon vertebrae in directing their overall therapy.

Although the success of massage in treating a number of disorders went virtually unchallenged by other practitioners, it failed to be integrated alongside *materia medica* in the standard medical school curricula. Most American and English physicians simply felt it was beneath them to administer treatment with their hands, let alone enter a field dominated by unorthodox healers. This attitude did not necessarily apply to other countries. "French, German, and Scandinavian physicians," Douglas Graham acidly remarked, "often apply massage themselves without any thought of compromising their dignity."[40]

The popularity of manipulation in these nations was largely the result of the efforts of Peter Henry Ling (1776–1839), a fencing master who combined body mechanics and gymnastics into what was popularly called "Swedish Movements." Ling's procedures were designed for both prophylactic and therapeutic purposes. His exercises were divided into active and passive movements, with the latter requiring the assistance of a trained specialist who would manipulate the patient through flexion and extension procedures. Though the Swedish medical community at first dismissed Ling's approach as being of no value, his methods nevertheless increasingly secured satisfactory results in cases where medication had been found wanting. As a consequence, a fair number of northern European physicians learned Ling's techniques and began applying them in cases of chronic, and even acute, disease. Hundreds of articles and many books on this system were published during the latter half of the nineteenth century.[41]

In examining the literature on massage and Swedish Movements, Still's new faculty recognized the similarity of the systems to osteopathy in the type of diseases successfully treated and in some of the techniques employed. However, they firmly believed that their own approach was more specific in terms of diagnosis and therapy. Charles Hazzard remarked:

Upon the whole these manual systems compare with osteopathy as does the shot gun with the rifle. They produce excellent results by the "shot gun method" of general manipulation, while osteopathy works with the definite aim of finding the obstruction to health and removing it. It is unavoidable that, if such a comparatively "hit and miss" method of massage can secure excellent results as a curative means, osteopathy, with its definiteness, must generally far exceed massage in results. It also follows that osteopathy must generally work more quickly and easily than massage in such cases as the latter could reach, and that it must succeed in a large class of cases beyond the power of these manual systems, since to this class belong so many disease conditions depending upon some removable obstruction not noticed by them.[42]

While emphasizing the supposed shortcomings of massage and Swedish Movements, Still's faculty were not loathe to borrow a number of those methods' underlying principles, particularly the importance of treating muscles and working manually to restore physiologic harmony in the absence of palpable anatomic displacement. In addition, advocates of these two systems provided them with experimental evidence of how manipulation had cured. Zubludowski, for example, found that massage increased electrical contractility of the muscles; Hopadze showed how it sped assimilation of food; Golz provided evidence that it aided the circulation of blood; Mitchell reported that it could produce an increase in red blood cells; and von Mosengeil found that manipulation promoted lymphatic absorption.[43]

The one area in which Swedish Movements and massage research could not materially assist the American School's faculty was in explaining why Still and his followers were obtaining their results by focusing predominately on the spine. To this question they found a partial answer in neurophysiology. In 1828 a Scottish physician, Thomas Brown, wrote an article in which he argued that pain about an internal organ could be caused by a disturbed vertebra that shared a common nerve supply. He called this phenomenon "spinal irritation."[44] In succeeding years this theory gained currency, and a number of books dealing with the subject were published.[45] One who accepted a similar principle but who did not use the term *spinal irritation* was the English surgeon and anatomist John Hilton (1804–78). In his popular and influential treatise *Rest and Pain*, first published in 1863, Hilton spoke instead of "sympathies," which covered the relationship between visceral pain without accompanying inflammation and "sore spots" about segmentally related vertebrae. To treat this type of pain, said Hilton, one must only treat the spine, which he did with rest and restriction of mobility.[46] While Still's staff seemed only vaguely aware of the doctrine of spinal irritation, they were quite familiar with Hilton's work and often cited his case studies in their lectures.[47]

Charles Hazzard and J. Martin Littlejohn argued by analogy that if referred pain could be produced by displaced vertebrae other remote symptoms, as Still argued, could be caused by them as well. Many of the nerves originating from the spine are connected to the sympathetic ganglia, whose function is to regulate blood flow to the various organs. Furthermore, many contemporary scientists speculated that the nerves had a trophic function—that is, they would directly supply nutrients to body tissues—so it followed that a disturbance of a spinal nerve could materially weaken the organ it supplied.[48]

The faculty also seized upon the principle of "stimulation and inhibition" as advanced by C. E. Brown-Sequard (1817–94). In animal experimentation he had discovered that a transverse sectioning of one lateral half of the base of the brain would be followed by augmentation of the motor properties in front of the cut and by inhibition of the opposite side. A stimulus weaker than normal would then be sufficient to produce an effect in the first case, while a stronger stimulus would be necessary in the latter. This meant that an "irritation" of a given nerve not only could reduce action at one distant part, it could also increase action in another.[49] Charles Hazzard, in his interpretation of Brown-Sequard's doctrine, believed that by putting physical pressure on "vaso-motor centers" along the spinal column the osteopath could return excessive or insufficient functional activity within an organ back to normal, independent of the actual cause. If, for example, a patient was suffering from a bad case of indigestion and there was no discoverable disturbance in segmentally related vertebrae, one could nevertheless relieve the condition by treating the relevant centers.[50]

In addition to finding scientific evidence supporting Still's theory, the faculty undertook the equally important task of making his ideas conform to established scientific facts, most notably the role of germs. Between 1876, when Robert Koch (1843–1910) isolated the bacteria responsible for anthrax, and the dawn of the twentieth century, the microorganisms causing fourteen different human afflictions were positively identified.[51] How could these discoveries be reconciled with the doctrine that anatomical misplacement was the major cause of disease? Similarly, what possible benefit could manipulating the spine have in treating infectious disorders? Still ignored the contradiction. "I believe but very little of the germ theory," he once declared, "and care much less."[52] All he seemed to admit to was the potential danger of germs in open wounds.

His faculty, however, preferred to face the problem more directly. Each of them accepted the existence and etiological role of microorganisms. Carl McConnell and the Littlejohns argued that while bacteriology seemed to undermine part of Still's original theory, its sister field, immunology, clearly supported him. Germs, they hypothesized, might be the active cause of disease, but spinal displacements, or what were then being called spinal "lesions," could be predisposing causes. If, as they believed, these structural lesions produced derangement of physiologic functions, it would follow that in their presence the body would automatically be put into a state of lowered resistance. Thus, correcting lesions shortly after they occurred would lessen the likelihood of germs' gaining a foothold in the body. By correcting lesions after infection had struck, the body's natural defenses could then more effectively respond to the invaders. Under these assumptions, osteopathic procedures seemed entirely applicable and necessary.[53]

Though he was at times disappointed and angry with his faculty because they sought to integrate the ideas of medical writers into their teaching, Still did not seriously interfere.[54] As a result, their contribution to the future course of the profession was assured. While his faculty's writings had no appreciable effect on the number of patients and students coming to the Missouri Mecca, they built osteopathy upon an intellectual base broader than the one Still was capable of constructing himself.[55]

IN THE FIELD

A lthough a few of Still's early graduates remained in Kirksville to serve as assistants in the infirmary, the majority went out into the field to establish their own private practices. A directory published in 1900 listing 717 graduates shows 121 (16.8 percent) residing in Missouri, 84 (11.7 percent) in Iowa, 83 (11.7 percent) in Illinois, 48 (6.6 percent) in Ohio, 32 (4.4 percent) in Pennsylvania, 31 (4.3 percent) in New York, and 30 (4.2 percent) each in Indiana and Tennessee, with the rest scattered throughout thirty-five other states and territories.[1] Some returned to their hometowns to begin work, while others were recruited by well-to-do patients to accompany them back to their city of origin to continue the treatment. Under such sponsorship the osteopath was formally introduced to the entire community.

ESTABLISHING A PRACTICE

The most important task for the freshly settled DO was to create a favorable impression on the townspeople. The system was new and in many areas unheard of. Often the term *osteopathy* was a handicap; quite a few prospective patients took it to mean that DOs thought all ailments were due to diseased bones or that they only treated fractures and dislocations. A few of Still's students recognized the potential problem and pleaded with him to change the name. He remained adamant. "I don't care what Greek scholars say," he bristled, "I want to call my boy osteopathy."[2]

In their advertisements in local papers or in printed brochures and journals, DOs explained that osteopathy was a totally original and independent system of health care. Several pointed out that it had "nothing in common with faith cure, Christian Science, spiritualism, hypnotism, magnetic healing, Swedish Movements, mental science, or massage."[3] Many in their audience, however, remained skeptical. All of those systems and others could involve, as did osteopathy, the "laying on of hands." Therese Cluett, DO, of Cleveland, found that this shared trait led to much misunderstanding:

> A lady entered my office and asked if I was a theosophist. I said, "No madam, I am an osteopathist." "Oh well," she replied, "It's all the same thing." Then it took me fully an hour to explain the difference between theosophy and osteopathy. On another occasion, I was approached with the question "Are you a Christian? because I don't want to take treatment from anyone who is not a Christian." This fairly caught my breath. . . . I asked her if she had put the same question to [her last physician] that she had put to me. She replied that she had not. It took me another hour to explain the difference between osteopathy and Christianity. For one patient I had to insulate the table, as they think this is some form of magnetic treatment. The next patient spies the insulators (as I had forgotten to remove them) and then there is trouble, as this patient won't have anything along that line of business.[4]

Cluett's problem was shared by Herbert Bernard, DO, who noted, "When I first came to Detroit, a woman telephoned me asking what price I charged to pray for people. Another one looked all over one of my operating tables trying to find the electric wires that he thought were hidden. . . . Quick results were dangerous in those days, as the patients would think there had been some rabbit's foot business worked upon them. They were afraid to tell of their relief . . . thinking people would take them for faith-cure followers."[5]

In explaining their system and differentiating it from others, many osteopaths told their patients that they alone could be considered "anatomical engineers." Only DOs knew where every bone, muscle, nerve, or blood vessel should be and what significance each held in the maintenance or restoration of health. Several of them published descriptions that were eloquently worded and simple to understand. If one accepted the metaphor of "man as a machine," the osteopath's logic made sense. As a violin or engine needed tuning or adjustment every so often, so also did the human body.

On the other hand, a good number of practitioners preferred the hardsell approach, which, though less dignified, was nonetheless successful in drawing attention. "Osteopathy," said one appeal, "deserves your patron-

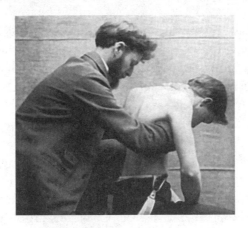

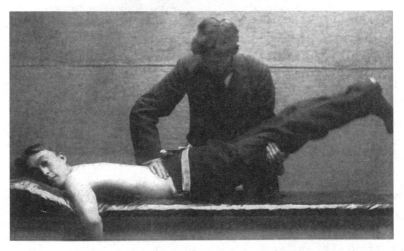

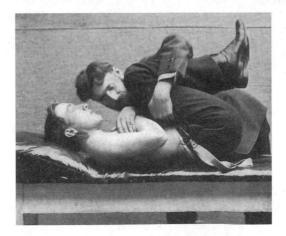

Dain Tasker, DO, of the Pacific School of Osteopathy, demonstrating
osteopathic manipulative techniques. Dain Tasker, *Principles
of Osteopathy* (Los Angeles, Baumgardt Publishing, 1903).

age because it has demonstrated its ability to do all medicine can do and much more. Many are the diseases entirely beyond the reach of the medical attendant that promptly surrender to the ability and the knowledge of the osteopath. In other words, there is not a single thing that medical men can surpass osteopaths in except . . . malpractice or killing people."[6]

Quite a few DOs published lists of the diseases that they claimed to be especially successful in treating. One typical advertisement included "headache, granulated eyelids, deafness, dripping eyes, dizziness, pterygium, polyps of the nose, cattarrh, constipation, torpid liver, gall stones, neuralgia of the stomach and bowels, dysentery, flux, piles, fistula, irregularities of the heart, kidney disease, female diseases, rheumatism and neuralgia of all parts, atrophy of the limbs, paralysis, locomotor ataxia, varicose veins, milk leg eczema, nervous prostration, hip joint disease, curvature of the spine, etc."[7] Some placed arbitrary recovery percentages next to disorders, such as one list, which reported these cure rates: sleeplessness, 95 percent; back pain, 90 percent; stomach trouble, 75 percent; dropsy, 65 percent; withered limbs, 60 percent; deafness, 55 percent; and cancer, 30 percent.[8] Others gave overall figures. "We cure about eighty-five percent of the cases we take," declared one infirmary, "benefit ninety-five percent, and fail on five percent."[9]

Another advertising method was the testimonial. Gratified patients would give the practitioner permission to publish flattering letters they had written. In defending this approach, the Matthews and Hook Infirmary declared, "Osteopathy is a great discovery. Its theory is most reasonable. But it has a practical side as well as a reasonable theory. And while it is perfectly proper to give its principles, its scientific basis, and speak in general terms of what it can do, it is also eminently necessary to take evidence and hear testimony from those who have tried it. . . . The question that the world asks is 'Does it work?' Osteopathy works. And for the benefit of those who wish to investigate we shall give from time to time the names and addresses of some who have thoroughly tested it."[10]

Most of the individuals throughout the country who patronized the early DOs suffered from chronic complaints similar to those found among patrons who came to Kirksville. "During our twenty months practice in Nashville," Dr. J. R. Schackleford noted, "we had many cases of interest, some of whom had gone the rounds of the medical profession, patent medicine, sanitariums, springs, mountains, sea shore, and various other devices and places for relief. Many who came said to us 'We have tried everything else and now we are willing to try osteopathy.' This is the rule in most cases, but whether we are the first or last makes but little differ-ence to us so [long as] we get the desired results."[11] Though some of the clinic reports sent into the *Journal of Osteopathy* concerned acute infectious disorders, these represented a minor portion of the average osteopathic workload. As W. L. Riggs, DO, wrote, "The idea is generally prevalent among the laity, wherever osteopathy is known, that the science is pecu-liarly adapted to long-standing and chronic cases, but that its results are too slow to counteract the rapid processes which follow the conditions prevailing in what are commonly called acute diseases."[12]

Patients were generally told that quick cures were the exception rather than the rule. Most DOs agreed with Dr. A. L. Evans, who observed:

> The over sanguine osteopath who advertises, writes, and talks constantly about cases that are remarkable for the rapidity with which they have yielded to the osteopathic treatment does himself and the profession an injustice. People are led to expect miracles. . . . It is wise to explain to them that it will take time to eliminate poisonous drugs from their system and to induce healthy normal ac-tion in torpid organs that have long been dependent upon extraneous stimula-tion. It is far better to impress this upon them than to tell wonderful stories— no matter how true—of marvelous cures effected in one or two treatments. By the latter method the patient is led to expect the same results in his own case and may be disappointed, for nature, though sure, is sometimes slow. If, on the other hand, more is accomplished than promised, osteopathy has won a friend that will never falter in allegiance to our system.[13]

To encourage this type of thinking, DOs generally billed their patients by the month, charging the standard fee for four weeks of treatment of $25. If the client's condition required an extended period, a sliding scale of charges was often worked out.

One problem generated by this arrangement was that patients expected as many sessions within the month as possible, regardless of their ailment. As a result, it became a matter of custom to administer no more than three treatments per week per client. Therese Cluett wrote of one supplicant who wanted "a treatment 'everyday' as Mrs. So-and-so goes to Dr. So-and-so and he gives a treatment 'everyday.' I say 'All right' knowing well it is only a question of time until she will beg off. In a week the patient is so prostrated by the frequent treatment that she is glad to admit she can-not stand so much osteopathy. It is all I can do to get her three times a week which is as much as anyone can stand without becoming debilitated."[14]

Since each of these encounters could last up to one hour, fatigue on the part of the patient as well as the practitioner can easily be imagined.

Of the early DOs in the field who contributed letters to the *Journal of Osteopathy* and other periodicals, almost all boasted that they were making a good living. In 1898 Joseph Sullivan, DO, declared, "Osteopathy in Chicago is on the high road to success. We are treating more people now than at any time during '97 and our results are most gratifying." Drs. F. W. and Mrs. Hannah noted, "Our patients now number three score of the leading people in Detroit and vicinity including representatives from almost every profession and avenue of business."[15] Drs. Mason Pressly and O. J. Snyder of Philadelphia claimed, "Within so short a time as a year . . . our books show that we are treating considerably over a hundred patients every month."[16] In explaining the osteopath's success, A. L. Evans listed several major factors: First, the theory of osteopathy was a rational and commonsense one; there was nothing "vague, mysterious, or occult about it." Second was the plain and reasonable plan of charges, "a system whereby the patient is enabled to tell approximately what it is going to cost him to regain his health." Third, manipulation was much more palatable to the patient than medicine or surgery: "If osteopathy did nothing but abolish experimental doses of poisonous drugs and curtail the number of blood operations it would be worthy of the gratitude of countless sufferers." And finally, Evans argued, "nothing succeeds like success. It is results that tell."[17]

Such missives of self-congratulation did not give a complete accounting of the situation generally for DOs, however. Each issue of the *Journal* would also contain notices by many DOs of their change of address, often from town to town. For them, osteopathy was not a sure-paying proposition, and not a few dropped out of practice altogether.[18] In many instances the business failure of the osteopath was due to public apathy; in others, an inability to impress his or her clientele was to blame. For some, it was a matter of the local MDs' employing existing medical licensing acts to drive them out before they had a chance to get settled in.

LEGAL STRUGGLES

The posture of orthodox physicians towards osteopaths varied considerably. Some regarded DOs as harmless quacks whose clienteles would patronize any new healer who happened to arrive in town. A few thought that their "rubbing" might be indicated in selected cases and would even refer an occasional patient or two. Most often an MD's response was shaped by the behavior of the DO. If the latter went quietly about his or her own business, there was usually small chance of confrontation. However, the osteopath who arrived in town with much fanfare, making extravagant claims regarding his or her own skill while intimating that the MDs were in league with the undertakers, was simply asking to be prosecuted. Whenever arrests did take place, the DOs would maintain that jealousy and fear were the prime motivating factors. Once they had begun to prove they were superior doctors, their argument ran, the MDs in self-defense would have to do all they could to get rid of them. While many did depart after being hauled before the courts, other osteopaths stayed to fight, and in the great majority of instances they managed to win.

The first legal action regarding a DO in the field appears to be the case of Charles Still (1865–1955), the founder's son, who had been invited to practice in Red Wing, Minnesota. When he arrived in 1893 he found himself in the midst of an epidemic of what had been diagnosed by local doctors as diphtheria. Though his experience to date had been with chronic disorders, Still was soon called upon to treat a victim. After his patient made a rapid recovery following conscientious applications of manipulative treatments to the neck, shoulders, and head, Still was asked to care for upwards of seventy children with reportedly only one fatality as a result. The State Board of Health, despite his apparent success, authorized his arrest for practicing without a license. By the time the case came up for trial Still's work had generated such considerable public support that the MD who had initiated the suit decided not to make an appearance, and the matter was dropped.[19]

Audrey Moore, DO, a graduate of the American School of Osteopathy's second class, was practicing in Macon, Illinois, when he was jailed on a similar charge. In his defense, Moore produced patients who testified that he had benefited or cured them when their MD had given up hope. "After examining a number of my witnesses," he recalled, "none of whom had seen any medicines used, and all of whom felt better after treatment, the justice said from the bench that all the people seemed to want to try this new humbug, so he would discharge the prisoner."[20]

A few DOs were even emboldened to initiate legal action against challenging MDs. In 1898 Harry Nelson, DO, who had been practicing in Louisville for about a year, became tired of the threats issued by the Kentucky Board of Health that he had better leave town or prepare himself for incarceration. In his suit, Nelson demanded that the board examine and license him or else cease and desist. When the matter came to trial Nelson's patients testified on his behalf; but unlike in the Moore case, the presiding judge was not impressed. Instead he listened to John McCormack, MD, of the American Medical Association, who maintained that "to license Dr. Nelson would be dangerous to the health, limbs, and lives of those citizens who might be treated by him in most instances." Though he had lost that round, Nelson would not give up his fight. The following year he brought his case to the Court of Appeals, which reversed the original decision and granted a permanent injunction against the board from preventing any DO from engaging in his profession. "So long as he confines himself to osteopathy, without the use of medicine or surgical appliances," the court ruled, "he violates no law and appellee should not molest him."[21]

What constituted the practice of medicine became the primary legal point at issue in most of the state courts that entertained such suits. MD representatives argued that the term *medicine* as found in the various healing arts statutes should be construed in its widest possible sense, while the DOs maintained that it meant the practice of administering drugs—and nothing more. In Alabama the state Supreme Court took the side of the MDs, deciding, "It is made entirely clear both by definitions and history that the word medicine has a technical meaning, is a technical art or science, and as a science the practitioners of it are not simply those who prescribe drugs, or other medical substances as remedial agents, but that it is broad enough to include all persons who diagnose disease and prescribe or apply any therapeutic agent for its cure."[22] However, only the Nebraska judiciary agreed. All other high courts ruling before 1904—Colorado, New York, North Carolina, Mississippi, Virginia, Ohio, and New Jersey—concurred with Kentucky that the term *medicine* should be narrowly interpreted.[23] "In forbidding an unlicensed person to apply any drug or medicine for remedial purposes," said the New Jersey high court, "the legislature plainly contemplated the use of something other than the natural facilities of the actor; some extraneous substance."[24]

In addition to their judicial struggle, both the MDs and the DOs traveled legislative avenues, appearing before a number of state legislatures to present their respective cases: the former sought specifically to outlaw the new system, while the latter wished to establish standards governing its practice. The DOs' first successful effort came in Vermont. Physicians in and nearby the town of Chelsea had become upset over the activities of Dr. George Helmer, who had established an osteopathic infirmary there in 1895. As Helmer's clientele grew the MDs complained to the state's attorney that the new healer was a public menace who preyed upon the weak-minded. Since several of the official's friends were among those being treated, their demands for prosecution were not looked upon favorably. This prompted the Vermont Medical Association to call upon the legislature for relief. Apprised of this, Helmer temporarily moved his offices to the capital to fight. While he was there, several lawmakers with chronic health problems decided to find out for themselves the relative merits of osteopathy by willingly submitting to his treatments. A number of them, including the lieutenant governor, were most pleased. As a consequence the legislature decided to throw out the medical society's bill, substituting and passing one giving any graduate of the American School of Osteopathy the right "to practice their art of healing in the state."[25]

Next in line to regulate the new system was North Dakota. Though DOs were involved in this lobbying effort, the battle was primarily waged by a patient, Helen DeLenderecie, the wife of "the merchant prince of Fargo." Her motivation was well expressed in a letter she wrote to the *Journal of Osteopathy*:

> In the fall of 1895, a lump appeared in my right breast. Our family physician advised its immediate removal assuring me that nothing but the knife could remedy the evil, and stating that it would soon assume a malignant form if not removed without delay. Knowing him to be a fine surgeon as well as a physician, I placed myself in his hands and submitted to an operation whereby my entire breast was removed. It was a great shock to my nervous system, and I had not recovered from it, when the same trouble appeared in my left breast. I had heard meantime of osteopathy and resolved to try it before again submitting to the knife. . . . I went to Kirksville and was completely cured in six weeks time. My own eyes saw and my own hands felt the obstruction that caused the trouble in both cases, and I knew very well that the knife was never necessary. . . . Osteopathy has clearly proven its right to recognition in the healing of cases heretofore declared only curable by the knife, and it is only right that its supporters should sustain its claim.[26]

When the bill came up for a vote in the Senate, DeLenderecie was given the unusual privilege of speaking to the entire body in its support. After hearing her dramatic story and her rebuttals of some of the arguments put forward by the MDs, the legislature passed the measure, and the governor, another osteopathic patient, happily signed it.[27]

In 1901 Mark Twain appeared before the New York State Assembly to speak on behalf of a bill legalizing osteopathy. To a gallery of cheering admirers Twain noted, "The State stands as a mighty Gibraltar clothed with power. It stands between me and my body and tells me what kind of doctor I must employ. . . . I know how Adam felt in the Garden of Eden about the prohibited apple. Adam didn't want the apple until he found out he couldn't have it, just as he would have wanted osteopathy if he couldn't have it."[28] In a private letter Twain wrote, "I want osteopathy to prosper; it is common sense, and scientific and cures a wider range of ailments than the [orthodox] doctor's methods can reach."[29]

The New York bill was not enacted and battles in several other jurisdictions were also hard fought. Nevertheless, in addition to Vermont and North Dakota, thirteen other states had by 1901 established laws regulating the new system—Missouri (1897), Michigan (1897), Iowa (1898), South Dakota (1899), Illinois (1899), Tennessee (1899), Montana (1901), Kansas (1901), California (1901), Indiana (1901), Nebraska (1901), Wisconsin (1901), and Connecticut (1901).[30] Many orthodox physicians had first thought osteopathy only a fad, but it became increasingly apparent to them that the actions of most courts and some legislatures were encouraging its growth and subsequent institutionalization.[31] At the turn of the century, when the American Medical Association was making considerable progress in eliminating the homeopathic and eclectic schools through a process of absorption, here was yet another competitor challenging the hegemony of orthodox medicine.[32]

OTHER SCHOOLS

While the fight in the courts and legislatures was in progress, a number of Still's graduates were forming their own colleges. The first were the National School of Osteopathy (1895) of Kansas City, the Pacific College of Osteopathy (1896) in Los Angeles, and the Northern Institute of Osteopathy of Minneapolis (1896). Within a few years the products of these schools, as well as of the American School of Osteopathy, established colleges in Boston, Philadelphia, San Francisco, Des Moines, Milwaukee, Chicago, Denver, and in smaller cities such as Wilkes-Barre, Pennsylvania; Ottawa, Kansas; Franklin, Kentucky; Fargo, North Dakota; Keokuk, Iowa; and Quincy, Illinois. Most of these institutions grew out of existing infirmaries where some clients, experiencing the benefits of osteopathy first hand, were anxious to become practitioners themselves. Instead of sending them to Kirksville, the proprietors, with an eye towards supplementing their income, were quite willing to organize their own programs. By 1904, of the estimated four thousand DOs in practice, approximately one-half were graduates of these other schools.[33]

The physical plants of these other colleges initially consisted of a small suite of rooms in an office building or a converted private residence. Since the first few classes were small, such facilities were seen by their operators as more than adequate. Entrance standards were nominal. While a number of catalogs called for a high school diploma, students lacking one found little difficulty gaining admittance, provided they were able to pay their fee in advance. Tuition was generally set at the American School of Osteopathy's original figure of $500 for the complete course, but because of competition it was soon lowered to a more reasonable $300–$350, which in turn increased the number of matriculants.

At first there were no common standards relating to the length and breadth of the curriculum. Some colleges, following the American School, limited their instruction to several months of anatomy, osteopathic diagnosis, and therapy, while others took it upon themselves to increase the time necessary for graduation as well as the number of subjects covered. Indeed, the Pacific College was the first to adopt a curriculum consisting of four terms of five months each which included broad basic science instruction.[34] When Still followed suit, months later, most of the other colleges decided to go along.

The teaching staff of these schools was small, generally between three and ten professors depending on the number of students enrolled. In some cases a majority of the instructors did not possess a DO degree or have any previous osteopathic training. MDs who wanted to learn something of osteophathic techniques as an adjunct to their own practice were pressed into teaching some subjects in lieu of part or the whole of their tuition fee. In almost all cases, MDs, whether they served on the faculty or not, were automatically given advanced standing, allowing them to complete the requirements for their diploma in approximately half the normal time.[35]

The equipment in these institutions varied markedly. Whereas the American School, the Des Moines School, and the Pacific College were able to move quickly into large, spacious facilities and furnish their laboratories with microscopes, dissecting and chemical analysis kits, and the newly invented x-ray machine, many others seem to have gotten along with a treatment table, a skeleton, and a few wall charts.[36]

In urging prospective students to enroll, each catalog made osteopathy appear to be a great calling and focused on the inner satisfaction one could expect by healing people in this "natural drugless way." However, if this was not sufficient motivation, there was always the appeal to one's mercenary interest. "The experience of graduates of osteopathy, who are now practicing in various parts of the country," claimed the Des Moines College, "demonstrates conclusively that there is no profession at this time in existence where a young man or woman can earn money so rapidly and successfully as in the practice of osteopathy. We have data in our office to show that good, scientific graduates of osteopathy can go out and earn from $250 to $500 per month, and in some cases their earnings reach as high as $800 a month."[37] The Northern Institute of Minneapolis was even more encouraging, declaring, "Osteopathy is the business opportunity of one's life time. There is increasing demand for it. No student properly equipped has made a failure of it. Individuals are making in cash from $500 to $1,000 per month. We know men who couldn't earn $1,000 a year who are now making $1,000 per month."[38]

Another argument ran that, whereas there might be as many as a dozen MDs in a small community, there would be a single DO who, after curing but a few of his or her counterparts' failures, would be swamped with business. An early catalog of the Philadelphia College observed, "There are not yet 400 osteopaths in the country, with a population of 75,000,000. The supply is short . . . the demand is great and there is no competition. This opens up a highway to success."[39] Correspondingly, the Atlantic School in Wilkes-Barre noted, "Fifty or one hundred years hence the profession will be crowded, but it will not be while we live. Those first in any field are the ones that reap the harvest—not the gleaners."[40]

Special appeals were directed at prospective female candidates. Since they were then denied entrance to all but a handful of regular medical schools, here was an alternative method of becoming a doctor. "The science of osteopathy appeals to women who desire a noble, uplifting work," the Pacific College reasoned. "A woman whose natural inclination is toward the benefit and assistance of the less fortunate of human kind, and who desires to allay herself with some work that while acting constantly as a moral uplift, will yet in an agreeable and rapid way place her peculiarly above all concern for the future, has the basis furnished her in osteopathy."[41] Such inducements were apparently quite successful, since approximately one-fifth of all graduates of osteopathic schools before 1910 were women.[42]

Each college naturally pointed to itself as the most advantageous institution in which to learn to become an osteopath. In addition to citing the alleged quality of their respective facilities, equipment, and staff, many focused on the environmental conditions of the city in which their school was located. "Franklin," said the Southern School, "is a noted health resort having several mineral wells whose properties are seldom excelled."[43] Similarly, the Pacific College declared Los Angeles to be "the best place in the world to study hard and maintain one's bodily vigor."[44] The Des Moines College even made a contrast between its town and the Missouri Mecca, claiming its "streets are well improved and the climate is exceptionally healthy," while Kirksville, because of supposedly poor sanitary conditions, "was rapidly becoming a hotbed for typhoid and malarial fever."[45]

Relations between these new colleges and the American School of Osteopathy were at best tepid and at worst openly hostile. Still believed that few if any of his early graduates had either the training or the practical experience to teach osteopathy on their own; that their institutions, for the most part, did not match the standards of the American School of Osteopathy; and finally, since some of them were situated within a few hundred miles of Kirksville, that they were in open competition for students who should rightfully be his.

The American School of Osteopathy declared war on the National School of Osteopathy in nearby Kansas City, Missouri, almost from its inception. The National School, headed by Elmer and Helen Barber, two graduates of the American School's second class, offered a regular course of instruction that was somewhat briefer than the one found at the parent institution, and it was rumored that its diplomas could be bought for a price. Elmer wrote the first book ever published on osteopathy, and in it he claimed that Still was wrong on a number of important theoretical issues and that anyone could learn to treat common ailments manipulatively with his text as the only necessary aid.[46] Not surprisingly, MD groups found Barber's work a most useful illustration of their contention that osteopathy was a fraud.

As a result of these goings-on, Still and his associates were placed on the defensive; they did all they could to dissociate themselves from and repudiate the Barbers, Elmer's book, and the National School. Dr. William Smith, who had entered practice for a brief time after teaching the first class and who had thus never met the twosome, was dispatched to Kansas City to determine whether the Barbers were complying with the new state law that required a college to give twenty months of personal instruction before awarding a diploma.[47] Meeting Elmer under an assumed name, Smith identified himself as an MD who knew all about osteopathy though he did not have the benefit of a DO degree. Barber, in turn, offered to issue him one on the spot for $150, a sum Smith agreed to and then paid. He next stopped at the attorney general's office, where he presented the bogus diploma and related the facts of the case, all of which led to the Barbers' indictment. Although the court found the pair guilty of violating the new statute, the judge refused to accede to the prosecutor's demand that their charter be revoked, finding that Smith's actual medical and osteopathic education mitigated the seriousness of the offense. After paying a small fine, the Barbers continued as before. Only in 1900, when their operation proved to be unprofitable, did they voluntarily decide to close their doors, but not before bestowing degrees on at least fifty individuals, some of whom established their own diploma mills, such as Noe's College of Osteopathy in San Francisco and Payne's College of Osteopathy and Optics in Ottawa, Kansas.[48]

Even more galling to Still than the Barbers' institution was the Columbian School of Osteopathy, located almost across the street from the American School and run by a former associate, Marcus Ward (1849–1929). Brought to Kirksville on a stretcher in 1890, Ward looked to Still for relief from a severe asthmatic condition. After he was restored to health, Ward entered into a business arrangement with his benefactor to learn his methods. Still later took him on as one of his assistants in the infirmary, and when Still established the American School of Osteopathy in 1892 Ward became a major stockholder and served as vice-president under the first charter. Within months after the college opened, however, the two had a falling out, and eventually Ward left town. He relocated in Ohio, enrolling in the medical department of the University of Cincinnati. After obtaining his MD degree there in 1897, Ward moved back to Kirksville. There, with the help of local businessmen who believed the town was large enough for two osteopathic institutions, he established the Columbian School.[49]

In his advertising Ward declared himself the "co-founder of osteopathy" and claimed to have been working along the same lines as Still since 1862, when he was thirteen years old. He also called himself the sole originator of what he named "True Osteopathy," which was the combination of *materia medica*, surgery, and manipulation. The use of all three therapeutic modalities, said Ward, would reestablish the "true" approach to healing as practiced by the ancient Greeks. Columbian students were therefore taught the principles of drug therapy along with other subjects now found in the expanded American School of Osteopathy curriculum. After they completed their twenty-month course and received the DO degree, they could enroll for another year of medical and surgical training, upon completion of which they would be granted the MD.[50]

Still did not address Ward's remarkable claim of being the co-discoverer of osteopathy. This he left to his friends and associates.[51] He did, however, sharply lash out at Ward's inclusion of *materia medica* in his curriculum. "Every man and woman sick and tired of drugs, opiates, stimulants, laxatives, and purgatives has turned with longing eyes to this rainbow of hope," he thundered, speaking of Kirksville, "and yet these medical osteopaths are trying to paint this rainbow with calomel and perfume it with whiskey." Ward's college, he opined, was a mongrel institution that, like the bat, is "neither bird nor beast." Anyone who pays his money into this institution, he claimed "gets neither medicine nor osteopathy, but a smattering, enough to make a first class quack."[52]

In its first two years of operation, the Columbian School attracted a fair number of matriculants; however, internal disputes between Ward and his financial backers would thereafter rack the college, and the institution closed in 1901 after graduating perhaps as many as seventy individuals. Once again the "co-founder" left town, settling in California, where for the next quarter-century he practiced in relative obscurity.[53] Though ostracized from the movement and quickly forgotten, Ward, with his efforts at fully integrating drug therapy into the osteopathic system, was a harbinger of battles to come.

STRUCTURE & FUNCTION

With the movement rapidly growing, many DOs thought it desirable to coordinate their efforts and activities. In February of 1897, a small group of American School of Osteopathy alumni met in Kirksville and decided to establish a national organization for this purpose. Graduates of other schools were then invited to take part in the planning, and by April they had collectively launched the American Association for the Advancement of Osteopathy, which was renamed and restructured as the American Osteopathic Association (AOA) four years later.[1]

The officers of the AOA under its 1901 constitution included a president, two vice-presidents, a secretary, and a treasurer, all chosen for a twelve-month period of service, and a board of trustees whose members were appointed for staggered three-year terms. Members of the board were charged with responsibility for the day-to-day affairs of the association, while the general membership—in reality only those opting to attend the annual convention—elected all officers, including the board, and decided questions of policy.[2] As the number of AOA members rose, this last feature of the system proved unwieldy, prompting those participating in the 1909 convention to enlarge the board from eleven to seventeen members and invest it with virtually complete control over policy issues.[3] In 1919 a dual form of central government was restored when the House of Delegates was created based on the proportional number of members within each state. These representatives, who were chosen by their respective divisional societies, thereafter selected all other national office holders and acted as the business body of the association during its annual week-long meetings.[4]

From its inception, the AOA actively worked to secure the conditions necessary for the movement to obtain professional recognition. To protect autonomy, it fought for independent boards of registration and examination. Academically, it both significantly lengthened the standard course of undergraduate training and supported ongoing research projects. To improve members' socioeconomic status, the association championed a code of ethics while combating the proliferation of impostors and imitators.

THE INDEPENDENT BOARD

At the turn of the century a majority of states were without a specific law governing osteopathy, and in several of the states with such an act, the legal position of the DO was hardly improved as a consequence. Early lobbying campaigns had usually been conducted by individuals speaking for only one segment of the emerging movement, leading to situations such as that in Vermont, which had extended practice rights only to graduates of the American School.[5] In other states diverse osteopathic factions had appeared before the legislatures with varying recommendations. This lack of unanimity often resulted in a poorly constructed compromise or no law at all.

Several of the early practice acts placed the regulation of osteopathy under the jurisdiction of existing state medical boards. In some states a DO was added to these agencies; in others no representation was granted. Although osteopaths would be examined alongside MDs, taking the same written tests in such subjects as anatomy, physiology, and chemistry, they were exempted from answering questions concerning *materia medica* or therapeutics. In a few states this arrangement seemed to work out satisfactorily for the DOs, as they found they could do nearly as well as the allopaths in passing examinations and becoming licensed. However, before other similarly constituted boards DOs did not have comparable success, and in certain instances MD officials prevented any comparison between the two groups whatsoever. In Iowa, for example, the legislature granted the medical board the power of accrediting osteopathic schools, with only graduates from approved institutions becoming eligible for licensure. After a cursory look at their catalogs, the board rejected all osteopathic colleges, thereby preventing any DO from legitimately practicing in the state and thus circumventing the intent of the lawmakers.[6]

In 1901 the AOA created a permanent committee to insure the passage of favorable laws. Toward this end, the Committee on Legislation devised a standard model bill for every state, whose chief feature was the establishment of independent boards of osteopathic examination and registration. In this plan each divisional association would nominate a long list of candidates from which the governor of the state would choose five to seven as members. These individuals, once they had been appointed, would be responsible for testing DO candidates, negotiating reciprocity agreements with other boards, and disciplining errant practitioners.[7] With the AOA trustees giving their strong backing to the idea, the committee began overseeing the lobbying efforts of the divisional societies. Frequently the societies faced a hard struggle. During the 1870s and 1880s, when medical practice acts were being reintroduced, several states granted the allopaths, homeopaths, and eclectics separate boards. This arrangement was sometimes difficult to administer and was often plagued with difficulties. Some states addressed these problems by abolishing this system and placing representatives of each sect upon a single, all-encompassing board, where they all kept a watchful eye on one another.[8]

In appealing to those who either were undecided about or saw no need for independent osteopathic boards, Arthur Hildreth, DO (1863–1941), the first chairman of the Committee on Legislation, hastened to argue:

> There has never been one single voice raised against osteopathy except by men of other medical schools. Every inch of progress made by our profession since its discovery has been contested by them. We have been looked down upon, criticized, ridiculed, called "faddists," "masseurs" and everything but gentlemen. And now when securing recognition by law, should we secure representation from existing Boards of Examination and Registration, we should have to do so against their protest and through the influence of our many, many, good friends. And after securing representation upon their boards, what is our position? Are we loved any more by them? No, we are still at a disadvantage because they overwhelm us in numbers and ours being unwelcome company, we need not expect many favors. Certainly we shall receive no help to reach out and grasp greater and better things such as must and will come to us with the right kind of encouragement and conditions.[9]

This line of reasoning became increasingly influential over the years, particularly where elected officials became convinced that discrimination by MD boards did in fact take place. In 1913, of the thirty-nine states that had passed osteopathic practice laws, seventeen provided for independent boards. Ten years later, these figures had risen to forty-six and twenty-seven respectively.[10] Furthermore, even in many of those states whose leg-

THE DOs: OSTEOPATHIC MEDICINE IN AMERICA

Early osteopathic view of medical politics.
From E. A. Booth, History of Osteopathy (1905).

islatures refused to accede to all DO demands, bills were enacted recognizing the AOA as the sole accrediting agency of osteopathic colleges, thereby preventing prejudicial actions by MD-dominated boards. Accordingly, the profession won for itself a considerable degree of autonomy and legal security.

LENGTHENING THE COURSE

The first group within the movement which attempted to set common educational requirements for DOs was the Associated Colleges of Osteopathy (ACO), founded in 1898 and composed of most of the legitimate schools. In fact, from the time of the ACO's formation, eligibility for membership in the AOA was predicated on being a graduate of an ACO-affiliated institution.[11] The Associated Colleges was created in part to remove the ill feelings the schools bore towards each other because of their aggressive competition for matriculants. Certain competitive activities—such as cutting tuition, stealing students, and shortening the time necessary to earn a diploma—were working to their mutual detriment. To stop these practices, each member of the ACO pledged to adhere to clear guidelines covering admissions, attendance, tuition, transfers, and advertising methods and to offer a mandatory two-year course.[12] Despite their promises, some of the colleges continued to engage in these prohibited practices, which only engendered further suspicion and distrust. As it became obvious that the schools could not effectively regulate themselves, the American Osteopathic Association in 1901 ruled that henceforth it would designate which colleges' alumni it would accept as members, in effect making the AOA the primary authority for establishing and maintaining academic standards.[13]

Looking at the state of osteopathic education at this juncture, leaders of the profession were convinced that major improvements were required. One of the critical areas of concern was the length of time needed to train and graduate DOs. Certainly the two-year curriculum of twenty months looked meager beside the four-year, thirty-six month program offered by almost all allopathic institutions. Wilfred Harris, DO, head of the Massachusetts school, argued, "The twenty month course is too brief. However clever the student, he cannot by any process of mental gymnastics, transplant himself with such suddenness from one field of thought and activity to another."[14] In 1902 the newly organized AOA Committee on Education issued a report urging the rapid establishment of a three-year course and the introduction of a four-year curriculum as soon as practicable.[15] According to the committee chairman, Dr. C. M. T. Hulett, this "would give time for more exhaustive work in many subjects now too much abridged; would make possible a substitution of the laboratory for the lecture, in many cases, and permit good laboratory work being made better."[16]

Not all DOs, however, saw matters in this light. Some believed that laboratory instruction was relatively unimportant. Others took the self-serving position that in adding one year and eventually another to the course, the profession would be declaring all previous graduates inferior or unqualified. This argument was skillfully answered by Dr. J. Martin Littlejohn, who, along with his brothers, had left Kirksville in 1900 to establish a school in Chicago. "The question is often asked did not our earlier graduates get along on much less time? Yes; but none have felt more than they the handicap that meant," he declared. "We do not mean they have not succeeded. They did succeed, but theirs was a struggle to evolve their knowledge as they advanced. To the busy practitioner, this is no easy matter."[17]

In 1903 the AOA and ACO jointly sponsored the first on-site survey of the schools. Chosen as inspector was Eamons Booth (1851–1934), who before becoming a DO, had earned a PhD from Wooster College and had taught at Washington University in St. Louis. In his report to the profession, Booth confirmed what others had already claimed in regard to the depth of preparation possible under the existing curriculum. His findings and recommendations helped to sway the undecided, and the AOA voted to require that all colleges inaugurate a compulsory three-year, twenty-seven month course by September 1904.[18]

A number of schools harbored great reservations concerning this policy, fearing a sudden drop in matriculants. Three early members of the ACO had recently folded—the Milwaukee College (1898–1901), the Northern Institute of Minneapolis (1896–1902), and the Northwestern College at Fargo (1898–1903)—primarily because of insufficient enrollment. There was concern that the new requirement might accelerate this trend. Curiously, the greatest objections were raised by the most solvent of the colleges. Charles Still claimed that all of his father's assets were tied up in the American School of Osteopathy and that in the event of its closure as a result of the proposed change, the "old doctor" would be ruined. The younger Still pleaded for an optional rather than mandatory three-year course, but the AOA rejected the request. However, they did by a narrow margin decide to give Kirksville an additional twelve-month grace period.[19]

While the total number of osteopathic matriculants markedly declined in the following decade, some schools were more dramatically affected than others. Closing their doors were the Colorado College of Denver (1897–1904); the Atlantic School, first of Wilkes-Barre, Pennsylvania, later of Buffalo, New York (1898–1905); the Southern School of Franklin, Kentucky (1898–1907); and the California College of San Francisco (1898–1910). In 1914 two other schools, the Los Angeles and the Pacific colleges, agreed to merge. By 1915 there were only seven recognized DO-granting schools in operation, located in Boston, Chicago, Des Moines, Kansas City, Kirksville, Los Angeles, and Philadelphia.

For most of the surviving institutions the addition of the third year had unexpectedly worked to improve their financial situation, as the decrease in new matriculants was more than offset by the extra year of tuition each student paid. This emboldened all of them to initiate an optional four-year course. In 1911 the Philadelphia school, spurred by recently enacted requirements for college registration in key states like New York, made the extra year compulsory for new matriculants.[20] It was soon joined by the Chicago College.[21] In 1914 the AOA Board of Trustees passed a resolution stipulating that the remainder of the colleges do the same no later than 1916.[22] Although some of the schools once again feared dire consequences because of this move, they realized that they had no choice but to comply. By 1920 all graduates of approved osteopathic colleges had received instruction equivalent in length to that of their MD counterparts.

SCIENTIFIC PUBLICATIONS AND RESEARCH

In 1901 the AOA introduced the *Journal of the American Osteopathic Association* (*JAOA*), whose purpose was to inform members of organizational business and to advance scientific knowledge. Within two years the *JAOA* had become a monthly publication of approximately fifty pages. Its staff recognized that for the *JAOA* to become a truly professional publication, the quality of its articles on practice, particularly those based on actual case histories, would have to rise above the level then prevalent. As one prominent DO succinctly remarked:

> It has long been appreciated by the public fully as well as ourselves, that osteopathic clinic reports in the true sense of the word DO NOT EXIST. What we call clinic reports and print in our magazines are a hodge podge of "hot air" and personal advertising in which we grant each other the right to advance rhetorically each his or her own personal reputation just as much as possible. . . . When [in] issue after issue our papers print glowing reports of what we have all done, and at that over our own signatures, isn't it just a little likely that the conscientious inquirer will say "Well do these people ever admit failures? Do they know what they fail to cure?"[23]

In order to improve the quality of osteopathic case reporting, the AOA Committee on Publication in 1902 appointed Edythe Ashmore, DO, of

Detroit, to lead a campaign in which practitioners in the field would be encouraged to fill out and submit concise patient histories, the best of which would be published in sets of one hundred as a semiannual *JAOA* supplement. This effort, it was thought, not only would be good experience for the average DO, but would also help to support osteopathic claims. Ashmore mailed out forms specifying the type of information needed, including client's age, sex, marital status, occupation, family history, prior treatment, symptoms, physical signs and diagnosis, what osteopathic lesions were present, the causes of disease other than lesions, and what urinalysis and other laboratory tests revealed. In terms of therapy, Ashmore requested descriptions of the specific manipulative technique employed, the length of the treatment, and changes in method as the case progressed.[24]

The first series was published in 1904, the last in 1909.[25] In each installment, cases were divided into eight broad disease classifications. Representation of given disorders did not always reflect the frequency of their appearance in a typical osteopathic practice; rather, many patient histories seem to have been selected on the basis of their value in demonstrating the alleged breadth of Still's approach. While most of these printed cases were described without the needless bluster, self-advertising, and harangues against the MDs, serious qualitative problems in the reports remained. Only a small number of examples where manipulation was found to be ineffective were included. Though these supplements were not meant for distribution to patients, most DOs had no desire to appear as anything less than successful before their peers either. Another difficulty was the lack of consistency in diagnostic findings. In a given condition, for example asthma, one DO would have found lesions along the cervical spine, another in the dorsal area, while a third would have located them in the lumbar region; and each would announce positive results by manipulating only where the lesions had been palpated.[26] These seemingly conflicting reports did not help the DOs refute charges by their MD critics that such lesions were imaginary and that osteopaths wrought their cures simply through suggestion.

As this weakness became manifest, influential DOs sounded a call for original scientific studies to "prove the lesion." In 1906 the AOA voted to establish and partially endow a separate institution to serve the dual function of conducting basic research and teaching advanced courses to DOs already in practice. Opposition to this plan was soon voiced by the colleges, several of which were already offering their own graduate-level classes and felt that the creation of a national center for this purpose would only lure away their students and fees. After three years of wrangling with the schools, the AOA agreed to drop the idea of a teaching role from their proposal, at which point financial contributions began to be solicited in earnest. By 1913 sufficient funds had been raised to purchase and equip a small building in Chicago which became known as the A. T. Still Research Institute.[27]

The first director of the institute was John Deason, DO (1874–1946), an American School of Osteopathy alumnus with an MS degree from Valparaiso University. He had also taken a postgraduate course at the University of Chicago, where he published a paper entitled "On the Pathways of the Bulbar Respiratory Impulses in the Spinal Cord" in the *American Journal of Physiology*.[28] Several of Deason's experiments and those of his associates centered on producing artificial "bony lesions" upon animal subjects and determining what effect, if any, they had on certain physiological functions. For the purpose of their research a bony lesion was defined as a slight dislocation or subluxation of a vertebra in relation to its adjoining segments. This was induced by manual adjustment of the subject under anesthesia and was verified immediately following and on regular intervals thereafter through digital palpation. In the first published compilation of their work, Deason and his colleagues recorded significant changes in carbohydrate metabolism, peristalsis, blood pressure, bile flow, and renal output following the artificial production of these lesions. However, their evidence supporting causal relationships was less than compelling.[29]

In 1917 a West Coast branch of the institute was established outside of Los Angeles and headed by Louisa Burns, DO (1868–1958), a 1903 graduate of the Pacific College who had later obtained an MS degree from the Borden Institute of Indiana.[30] When Deason left basic research for private practice during the First World War, Burns emerged as the profession's only full-time investigator. Her experiments were similar to those that had been carried out by the Chicago group, with some modifications. In her long career, Burns wrote several books and monographs in which she claimed that a variety of functional and organic disturbances of the eyes, heart, lungs, kidneys, stomach, and other viscera in laboratory animals were directly attributable to artificially produced lesions.[31] Although she never published in outside science journals, many of her DO contemporaries were convinced that her internally financed studies demonstrated the soundness of their system. However, Burns failed to provide adequate controls, and her conclusions were not consistently derived from the data she presented.[32] As a consequence of these inadequacies in her accounts, and because little other research into the basic science of osteopathy was being carried out at the colleges prior to the Second World War, fundamental questions concerning the etiology and role of the lesion in disease remained unsatisfactorily answered.

THE CODE OF ETHICS

In the early days of the movement, rarely did one osteopath locate his practice near that of another, except in major cities. However, as the number of new DOs increased, it became common for two or more to serve a relatively small town. In such communities, particularly where osteopaths and allopaths competed for limited health dollars, price wars and instances of character assassination took place. These occurrences made osteopathy appear as something other than a lofty calling—an impression furthered by those DOs who engaged in indiscriminate advertising.[33]

In 1904 the AOA adopted a formal code of ethics establishing guidelines for proper professional conduct. This document, based in part upon the code of the American Medical Association, emphasized cooperation rather than competition. To eliminate price wars, all DOs in a given geographical area were encouraged to formulate definite rules governing "the minimum pecuniary acknowledgment from their patients."[34] This concept was not unheard of within the ranks prior to establishment of the code. Members of the Washington State Osteopathic Association had agreed two years earlier to abide by a uniform fee schedule, charging no less than $2.00 for single office visits, $2.50 for single house calls, and $3.00 for single night visits. Chronic cases were billed at $25.00 for the first month, $20.00 for the second, and $15.00 for each subsequent one. Ministers and schoolteachers received special reduced rates, while the poor were to be treated for free.[35] With the AOA now behind this type of arrangement and the Washington plan working to the participants' satisfaction, several other divisional and local societies devised their own schedules.

The code of ethics also prohibited DOs from pirating one another's clients, declaring:

The physician, in his intercourse with a patient under the care of another physician, should observe the strictest caution and reserve, should give no disingenuous hints relative to the nature and treatment of the patient's disorder, nor should his conduct directly or indirectly tend to diminish the trust reposed in the attending physician. . . . A physician ought not to take care of or treat a patient who has recently been under the care of another osteopathic physician, in the same illness, except in the case of a sudden emergency, or in consultation with the physician previously in attendance or when that physician has relinquished the case or has been dismissed in due form.[36]

Significantly, the code was ambiguous on whether this last courtesy was to be extended to MDs.

Unethical advertising was also denounced in this document, and later the AOA published a list of what it found to be the most offensive practices. These included buying newspaper space, publishing field literature that contained a "percentage of cures," and issuing statements the truth of which was open to legal question.[37] The association did not frown on all advertising, however. One type of promotion which was looked upon with great favor was the lay-oriented osteopathic health journal, such as the one established by Dr. Henry Stanhope Bunting (1869–1948). Working as a reporter for a Chicago newspaper, Bunting was sent off to Kirksville in the mid-1890s to write a story on Still and his movement. Impressed with what he found, he soon returned to enroll. After graduating with his DO degree in 1900, he settled again in Chicago, where he started a practice and took night classes at a medical college to further his education. In 1901 the busy Dr. Bunting introduced two continuing monthly publications; the *Osteopathic Physician*, for the practitioner only, dedicated to voicing all sides of every professional controversy; and *Osteopathic Health*, which was aimed exclusively at the general public. Compared to previous lay literature, *OH*, as it was commonly called, contained little in the hard-sell vein. Instead, there were broad discussions of the philosophy, principles, and practice of osteopathy. Bunting, who maintained an avid interest in advertising theory and wrote a textbook on the subject, believed that the most effective means of getting the attention of people was via the underplayed message.[38] Needless to say, this meant a more dignified approach. A DO in the field could send Bunting a list of names and addresses of actual or potential patients, and for a standard fee Bunting would notify those so-designated that they would receive a one-year subscription to *OH*, into each issue of which he would insert a professional card of the practitioner paying for the service. As this system became popular, the AOA in 1914 decided to publish its own lay vehicle, the

Osteopathic Magazine, which also included general articles that were not health-related. As each of these and other new advertising ventures demonstrated their value in generating new business, the desire for and use of more questionable methods greatly diminished.

For those members of the AOA who were unwilling to abide voluntarily by the provisions of the code of ethics, disciplinary action became necessary. Every year the Board of Trustees investigated alleged misconduct, suspending or expelling those found guilty of serious violations from the ranks of the association. However, not all osteopaths sought membership in the AOA. In 1918 only 51 percent of the approximately 6,000 DOs belonged. In 1930, 57 percent of roughly 7,600 practitioners were in the fold.[39] Thus, for several decades almost one-half the total number of DOs were outside the influence or control of the AOA. This jurisdictional gap was filled to some extent by the state osteopathic or medical boards of registration and examination, which had the power to revoke licenses for a variety of reasons coming under the heading of unprofessional conduct. Therefore, although instances of disreputable behavior would continue to be a problem for the movement, organized osteopathy had established the basic institutional mechanisms for dealing with unethical practitioners.

IMPOSTORS AND IMITATORS

By the turn of the century, correspondence schools teaching osteopathy were springing up around the country, particularly where there were no osteopathic practice laws yet in force. In Ohio, for example, a man claiming to be an MD as well as a DO offered a teach-yourself-at-home textbook and a handsome diploma, both of which could be purchased for only $25. In New York, where the cost of living was considerably higher, a Norwegian ex-sailor announced a similar service for $100.[40] The number of bogus osteopaths thereby produced can only be guessed at; nevertheless, their impact was undeniable. S. C. Matthews, a DO in Wilkes-Barre, Pennsylvania, complained, "There are towns within my knowledge where disreputable and bungling methods of the unauthorized and uneducated practitioner have so injured the name of our science, that a legitimate osteopath would have the utmost difficulty in establishing himself. At best, it would be a struggle of many weary months."[41]

Since the correspondence schools depended upon newspapers and magazines to attract their "students," the Committee on Education, to whom the AOA Board of Trustees assigned the task of closing them down, decided they would first focus their efforts on the periodicals themselves. The committee reasoned that if publishers were made aware of the absurdity of these charlatans' claims, they would refuse to carry their messages. It sent a standard letter that read in part: "Would you accept the advertisement of an institution which offered to fit persons for the practice of medicine by a correspondence course of study? Yet it is just as impossible to fit a person by mail for the practice of osteopathy."[42] So that publishers could better appreciate the situation, the committee attached to their plea a description of the minimum requirements a college needed to obtain AOA approval, plus an abstract of existing state statutes. Most of those so contacted wrote back that they would henceforth reject such ads, and in 1907 the committee reported to the AOA Board of Trustees that there remained only one magazine of any sizable circulation that refused to honor their request.[43] Though the selling of mail-order diplomas did not end as a result of the committee's actions, it ceased to be a critical issue for the profession, especially as osteopathic legislation grew more widespread and those practitioners with unearned degrees became subject to prosecution under the law.

Quite a different problem, however, was presented by those individuals practicing what appeared to many to be osteopathy under a different name. The most numerous of these were the exponents of chiropractic, founded by Daniel David Palmer (1845–1913). According to Palmer the principles of this system were fashioned by him in 1895, while he was making a living as a magnetic healer in Davenport, Iowa. A janitor who worked in the building where Palmer kept an office told the practitioner that he had gone deaf seventeen years earlier after something "gave way" in his back. Reasoning that a displaced vertebra was responsible, Palmer manipulated the spinal segment into its proper position, and the janitor announced that his hearing had returned. Based on this and subsequent cases so treated, Palmer declared that 95 percent of all disease was due to "subluxated" vertebrae.[44]

In 1898 Palmer began to teach his methods. Initially he found few followers, training only 15 students through 1902. Business picked up for a while, but then Palmer's personal good fortune declined. In 1906 he was convicted of practicing medicine without a license and was sentenced to spend six months in jail. During his incarceration, his school was taken over by his son, Bartlett Joshua Palmer (1881–1961). The two were better known as BJ and DD. When DD was released, BJ squeezed him out of the college, whereupon DD tried without success to operate schools elsewhere. Returning to private practice, the elder Palmer wrote a massive textbook, a significant portion of which was devoted to a diatribe against his son. Bitter feelings between the two remained strong. At a founder's day parade held in Davenport in August of 1913, the uninvited DD, marching on foot, was struck from behind by an auto driven by BJ. DD died a few months later, with some of his followers convinced that his death was a consequence of his injuries.[45]

Under the younger Palmer the school continued to grow, securing many matriculants by sensational advertising—a practice BJ encouraged his followers to emulate. By 1916 there reportedly were some fourteen hundred students in attendance, taking one year's training leading to a doctorate in chiropractic, or DC, degree. For those who could not appear in person, a correspondence course was instituted. As the Davenport college flourished, dozens of other chiropractic schools, the great majority of them engaged in the selling of diplomas, were established across the country.[46]

Many early chiropractors were arrested on the charge of practicing osteopathy without a license. Unlike those with fake DO diplomas, however, chiropractors claimed that they were not pretending to be osteopaths and were therefore innocent of any offense. In court they cited a number of differences between the two systems. The DOs, they pointed out, commonly adjusted several vertebrae to treat a given disorder; they invariably adjusted but one. The technique also varied. Osteopathic manipulations were based on the lever principle, namely, the application of pressure on one part of the body to overcome resistance in motion elsewhere. This meant twisting the patient's torso in certain directions while maintaining a steady hold upon the point to be influenced.

The most common chiropractic procedure of the era had the client lying prone with little, if any support below the spine. The operator would then place both hands directly over a vertebral segment that was believed to be "subluxated" and administered a quick thrust downward with all possible force. In court, when DO witnesses were called to the stand, they would often testify that this method was crude and dangerous and would not be employed in osteopathic practice. Such statements, however, unintentionally worked to the chiropractors' advantage, since they indicated to juries that there were indeed divergences in approach. With respect to the element of danger, the defendants were only too glad to present patients who had been so treated, attesting to the safety of such maneuvers.

To further cement their position, some chiropractors cleverly managed to obtain and circulate signed letters by officials of recognized DO-granting schools stating that a course of chiropractic was not the same as one in osteopathy. As a result of these tactics, they generally won acquittal.[47]

Since the courts were beginning to establish the chiropractors' right to engage in their livelihood outside the jurisdiction of either the medical or osteopathic licensure acts, several legislatures realized that unless they passed laws recognizing the group, their states would be inundated with diploma mill graduates. In 1913, despite vigorous lobbying of MDs and DOs alike, Kansas and Arkansas became the first to enact chiropractic bills. Each required for licensure an eighteen-month course of personal instruction at a duly chartered college. By 1922 twenty other states had similar statutes.[48] At this time the number of DCs legally and illegally in practice probably exceeded the number of osteopaths in the country. Thus, while the DOs, through the AOA, had made considerable progress in obtaining some professional recognition insofar as certain measures of organization, autonomy, socioeconomic status, and education were concerned, they nevertheless could not prevent the rise of others who could more inexpensively and quickly produce practitioners capitalizing upon the therapeutic modality that was the central feature of the osteopathic system.[49]

5

EXPANDING THE SCOPE

The most controversial issue the DOs wrestled with throughout the first three decades of the twentieth century was the scope of their practice, particularly in regard to the range of therapeutic modalities they should utilize and the type of diseases and conditions they should treat. Vying for the support of the majority of practitioners were two distinct groups. One was composed of the self-proclaimed "lesion osteopaths." In their view, Still's system consisted of structural diagnosis and manipulative therapy. They felt that the only thing necessary to do for the patients they saw was find lesions along the spine or elsewhere and proceed to eliminate them. This, after all, was the same approach that Still had successfully employed to permit crippled people to throw away their crutches and other chronically ill individuals to lead a more normal life. Opposing this group were the so-called "broad osteopaths." While these DOs strongly believed in the efficacy of manipulation per se, they were not willing to limit themselves, envisioning the osteopath's role as that of a complete physician able to deal with any case and using whatever means were needed to best help the patient.

SURGERY AND OBSTETRICS

The first open debate between the proponents of lesion and broad osteopathy arose over whether or not the DO should receive an education in and practice surgery and obstetrics. Lesion osteopaths argued against the inclusion of these areas of practice, principally on the grounds that the DO could not be expected to perform two or more different roles as well as one. Why scatter one's energies and attention to other disciplines, no matter how intrinsically worthwhile? If patients were in need of a surgeon or an accoucheur, they could easily be referred to an MD specialist. The broad osteopath saw this reasoning as short-sighted. They maintained that if osteopathy was to rank with allopathy, homeopathy, and eclecticism, it was imperative that it provide the same range of services to its clients as they did.

Since the American School of Osteopathy's curriculum did not at first encompass any training in surgery or obstetrics, many early lesionists assumed that their position was in conformity with Still's. However, available evidence strongly suggests that he originally wanted to add these subjects,[1] and once they were integrated in 1897 he gave them his full support. In 1901 he wrote that his students were to be taught all operative surgery commonly performed in rural areas and were to become knowledgeable in the handling of obstetrical cases. "In short," he declared, "our school is prepared and intended to qualify its graduates when called in counsel or to lead that they might have the necessary information at that time so they will not be handicapped or embarrassed."[2]

The American School of Osteopathy having led the way, the other colleges followed suit. At the time Booth undertook his survey in 1903 all the institutions he visited were conducting classes in obstetrics and gynecology and in surgery.[3] When the length of the curriculum was increased, so too were the listed catalog hours devoted to these courses. For the 1908–9 academic year, the Kirksville, Chicago, Philadelphia, and Los Angeles colleges offered a combined average of 293 hours of instruction in these areas. By 1918–19, this figure had risen to 802.[4] As this trend became clear, the lesionists gradually, if somewhat reluctantly, came to accept these subjects as legitimate features of osteopathic practice.

Before 1920 comparatively few DOs performed surgery other than setting fractures and closing minor wounds. The paucity of opportunities for education in surgery, combined with restrictive state licensing laws for osteopathic surgeons, made this area of practice relatively unattractive. Significantly, DO students who expressed a desire to become surgeons were encouraged by their teachers to obtain a valid MD degree once they graduated so they could receive the depth of instruction required and not be legally circumscribed.[5] Obstetrics also constituted a relatively small fraction of the typical osteopath's workload. In 1917 one DO took a survey of his colleagues and found that while 52 percent of those sampled were accepting obstetrical cases, the average practitioner who did handled fewer than five deliveries each year.[6]

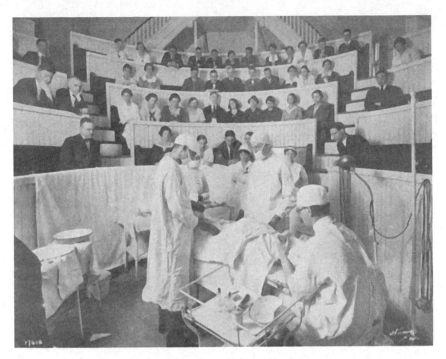

Early osteopathic instruction in surgery. *Courtesy of Philadelphia College of Osteopathic Medicine.*

A few osteopaths in this era did specialize in surgery or obstetrics, and most of them employed manipulative therapy in their practice, believing that this modality gave them a decided advantage over their MD counterparts. In obstetrics, strictly osteopathic procedures were thought to shorten the duration of labor, lessen the pain, prevent mastitis, and secure a more rapid convalescence of the patient.[7] In 1912 Lillian Whiting, DO, of Los Angeles, published data showing that of ninety-nine primiparae cases who received one to seven months of manipulative treatment prior to birth, the average duration of labor was 9 hours and 54 minutes, compared to 21 hours and 6 minutes for twenty-four untreated clients. She noted similar differences in multiparae deliveries.[8]

In surgery, Harry L. Collins, DO, MD, of the Chicago College claimed that four distinct benefits were to be derived from the osteopathic approach: first, fewer patients needed operations; second, when surgery was indicated the work involved was less extensive than expected of similar cases that had not received previous osteopathic care; third, a DO surgeon thoroughly grounded in osteopathic principles was less apt to sacrifice tissue needlessly; and fourth, the patient ran the postoperative course more smoothly and encountered fewer complications.[9] Of these assertions, the last was given special emphasis. About 1911 George Still, DO, MD (1882–1922), of Kirksville, the founder's grandnephew, who had earned his medical diploma and a master's degree in surgery from Northwestern University, began administering manipulative therapy to his surgical patients after they had undergone operations. His working theory was that this treatment would prevent blood stasis and speed lymphatic absorption, an approach that would aid the body's natural defenses against infection. Soon after Still instituted this protocol, a dramatic decline in the rate of postsurgical pneumonia was recorded among his patients. Indeed, he was so satisfied with these results that he decided to forswear the common practice of giving strychnine after surgery as a means of stimulating the heart. This omission seemed only to increase the overall benefits. He told his colleagues, "In our post-operative cases, study the charts and you will see that they do not have the acutely violent cases that usually occur under other treatment. . . . Instead of having a temperature of 105, pulse 165, respiration 70 . . . they are more apt to run a temperature of 102, pulse 120, respiration 35 to 40."[10]

At the same time that osteopathic surgeons were broadening the possible applications of manipulative therapy, they were also pointing out when such treatment was contraindicated. In 1904 Frank Young, DO, MD, then at the American School, wrote a textbook entitled *Surgery from an Osteopathic Standpoint* in which he cautioned against manipulation of patients with ankylosis, dermatitis, hernia, skin ulcers, glanders, cysts, osteomyelitis, scurvy, gangrene, and septicemia. In succeeding years others expanded the list. S. L. Taylor, DO, MD, president and surgeon-in-chief at the Des Moines College, observed that five common disorders that osteopaths were treating manipulatively—inflamed tonsils, hemorrhoids, fibroid tumors, gallstones, and appendicitis—were often more successfully handled by the scalpel. James Littlejohn, DO, MD, of Chicago argued that in gynecological cases the presence of pustulant inflammatory processes, new growths, displacements, congenital defects, and traumatism signaled surgical and not manipulative intervention, and Dr. George Still chastised osteopaths who adjusted the spine in Pott's disease.[11] As a result of these warnings, DOs became more cognizant of some of the limitations and possible hazards of the founder's methods.

With the addition of surgery and obstetrics to the curriculum, A. T. Still believed his system to be complete. His graduates could deal with a wide range of ailments and conditions and could utilize all the modalities he felt essential to general practice. The broad osteopaths, however, were not satisfied, looking with favor upon additional drugless tools such as hydrotherapy, suggestive therapeutics, and electrotherapy, which seemed of value in certain cases. To Still and the lesionists, the employment of these "adjuncts" constituted heresy. In 1902 Still made his feelings known in "Our Platform," an unsigned manifesto consisting of nine campaign planks and offered by him, through publication in the *Journal of Osteopathy*, as the official view of the profession. "The fundamental principles of osteopathy," Still declared, "are different from those of any other system and the cause of disease is considered from one standpoint, viz.: disease is the result of anatomical abnormalities followed by physiological discord. To cure disease the abnormal parts must be adjusted to the normal, therefore other methods that are entirely different in principle have no place in the osteopathic system."[12]

Responding to this platform, though not to Still personally, was Dain Tasker, DO (1872–1964), a graduate of the Pacific College who at the time was completing a book on osteopathic principles which became a standard text in the schools. Of the founder's view of the etiology of disease, Tasker wrote: "This may be the sum of some people's osteopathy, but it is not mine. I would really like to know how many men of five years active practice are willing to balance themselves on this two-inch strip of a plank. . . . I doubt whether a man who is satisfied with it could be convinced by any line of reasoning whatsoever that life in its manifold phenomena has any other side than the mechanical. . . . FUNCTION DOES AFFECT STRUCTURE JUST AS DECIDEDLY AS STRUCTURE AFFECTS FUNCTION." Turning his attention to adjuncts, Tasker was also direct. "There is no reason," he noted, "why each member of our profession should not feel free to develop and fit himself to aid humanity by the use of sunlight, X-radiance, hydrotherapy or any other method which appeals to his best judgment. . . . In order to be truly scientific we must love truth better than we love our preconceived ideas of what truth is."[13]

The following year, at the 1903 AOA Convention in Cleveland, this controversy came to a head when Dr. William Smith took the floor and began ridiculing some of his colleagues for a few of the modalities they were using. "I gave up medical practice and why?" he rhetorically asked. "Because I thought I got something better. . . . And so today when I look around me and I see so many adjuncts to osteopathy, when I find this man using the colon tube, and the other man using the vibrator to treat the eyes, and a third using electric massage to fix up a patient's back, another man with a static apparatus to restore manhood . . . and another with something to grow hair on bald heads, I ask you where in the name of common sense is osteopathy in all that?" Smith was immediately seconded by Herbert Bernard, DO, of Detroit, who observed, "At this stage of osteopathic history, when there is so little known and so much to learn, is it not foolish to tie to the osteopathic kite a tail made up of electrotherapeutics, hydrotherapy, with a few other adjunct knots tied in it? People in looking at it from a distance might mistake the tail for a kite. They, the people, are very likely to call osteopathy anything else but what it is anyway. Is it, can it be possible, that some of us are helping them to do this?"[14]

The defenders of the adjuncts in turn claimed that their opponents gave manipulation too much credit. C. W. Young, DO, a graduate of the Northern School, noted, "Dr. Smith spoke of the use of the colon tube. I have interviewed a number of Kirksville graduates in Minneapolis and St. Paul and I have listened to their talk pertaining to this matter but I have never yet learned any purely manipulative method which will invariably move feces in the colon. And if it was one who was near and dear to me above everything else that Dr. Smith was called to treat, and some hot water and the colon tube would save that life, and he refrained from using them in order to stand by osteopathy, I would not think of him as being much less than a murderer." Responding to Young's harsh attack on Smith, C. M. T. Hulett, DO, caustically implied that the former's education left something to be desired: "Now then, Dr. Young never studied under Dr. Still. He got his osteopathy second hand. It may be just as good, but when he asserts that osteopathy as taught by Dr. Still is deficient, he must prove that Dr. Still and those men [he instructed] failed, not that he failed in order to substantiate his position."[15]

Though this debate was largely fought between graduates of the American School of Osteopathy and the alumni of other schools, many of which had already integrated one or more adjuncts into their curriculum, the Kirksville group was by no means of one mind. Dr. Carl McConnell, the author of a major manual on osteopathic practice, in which he attempted to reconcile Still's ideas with those of the distinguished orthodox physician William Osler, had declared on a previous occasion:

> While I am perfectly willing to concede the major part of what our therapeutics should be to manipulation, I am by no means willing to assert that every disease or ailment of the body means "readjustment" of certain tissues, in order to restore health. I have heard one or two argue that it makes no difference whatever one uses as food provided his vertebrae, ribs, etc. are in correct position. It would be quite laughable, if it were not so serious to hear such narrowmindedness. If their proposition were true, medical knowledge prior to the discovery of osteopathy amounted to naught. They do not seem to realize that it was through medical knowledge already existing that osteopathy was developed. It is just such people as this that harm us more before the medical world and public more than anything else. They will bring up their "manipulative argument" when they do not have the first conception of hygiene, preventative medicine, etc.[16]

With opinion sharply divided, no consensus was obtained on the issue of adjuncts at the convention. Given a lack of policy directive, it was left to individual schools and practitioners to decide their own course. In the ensuing years the lesionists were encouraged by the discarding or avoidance of some adjuncts by most of their colleagues; however, sentiment had clearly shifted towards the position of the broad osteopaths on others.[17] Indeed, certain agencies became so acceptable—hydrotherapy, corrective exercise, diet and food chemistry, and mental therapeutics—that in 1912 the *JAOA* introduced monthly columns on each. Although these and other drugless modalities came to occupy only a minor place in the college curricula and in patient management, their integration was nevertheless important, since it marked the first significant divergence by the majority of the profession from the original doctrines and charismatic authority of Still.

CHEMICAL AND BIOLOGICAL AGENTS

While the questions of whether surgery, obstetrics, and the so-called adjuncts should become part of the osteopathic system resolved in a few years, the issue of chemicals, vaccines, serums, and endocrines followed a more difficult path to resolution. Despite that the same principles put forward in the adjuncts controversy were relevant here—the freedom of the practitioner to choose any modality thought helpful in the management of a given disorder and the right of the schools to teach what they desired—the symbolic meaning of chemical and biological agents to the pioneers of the movement made this a lengthier and more painful matter to settle.

The last third of the nineteenth century was marked by several momentous changes in the practice of orthodox medicine. With each passing year an increasing percentage of regular physicians came to rely on a smaller number of drugs in less heroic doses for those conditions in which pharmaceuticals seemed indicated.[18] Though as late as 1899 the *Merck Manual of Therapeutics* listed sixty-eight different treatments for diabetes mellitus—including arsenic "for thin subjects," codeine ("a most effective remedy sometimes requires to be pushed to the extent of 10 grains per day"), iron ("most useful with morphine"), and belladonna "full doses"— the great majority of younger American MDs were rejecting this empirical approach.[19] Medical thought was also transformed by the emerging fields of bacteriology and immunology and the work of such scientists as Louis Pasteur, Robert Koch, Emil Behring, and Paul Erlich, who shifted the focus of practice from eliminating the symptoms of infection to destroying or rendering inert pathogenic microorganisms and their byproducts. This was made possible largely through vaccines, which allowed sufferers to manufacture their own antibodies, and serums, which already contained the specific antibodies of another human or animal.[20] By 1900 scientists had developed prophylactic and therapeutic agencies for rabies (1885), diphtheria (1891), tetanus (1891), cholera (1892), plague (1897), and typhoid fever (1898).

A. T. Still, for the most part, was unimpressed by these advances, believing that the chemical and biological tools employed by the MDs were often toxic to the body, as well as being vile and disgusting.[21] Furthermore, the regular physicians were ignoring the structural basis of disease. One might conceivably eliminate symptoms through such modalities, but not the underlying cause, namely, structural lesions. Finally, whatever the alleged usefulness of these agents, osteopathy was always equal to the task. Still claimed, for example, that he could prevent the chills and fever of malaria without quinine by periodically adjusting the lumbar vertebrae; disperse the fluid in dropsy without digitalis by treating the eleventh and twelfth ribs; and reduce the swelling of a gouty big toe without colchicine by manipulating the foot. As for the diphtheria antitoxin, his son Charles had proven in Red Wing, Minnesota, that a DO could get spectacular results without it.[22] The only orthodox medicinal agencies Still did sanction were anesthetics and antiseptics in surgical and obstetrical practice, and antidotes in poisoning cases.[23]

The first DOs who were in favor of adopting a wider variety of pharmaceuticals, vaccines, serums, and endocrines, were those who also held a medical degree. Their additional training and experience had convinced them that some of these tools had proven their worth and there was no valid reason for not using them in patient management. If a client was suffering from gout, it made sense to them to both manipulate and administer colchicine. Similarly, in malaria, why not adjust the spine and give quinine together? In this fashion, the sufferer would receive the best of both systems.

Still did not think much of this view. In 1903 he rhetorically asked,

> What will become of the MD DO? He ought to be put it a class by himself and no doubt will be if he attempts to practice osteopathy and medicine combined. . . . Medicine and osteopathy as therapeutical agencies have nothing in common either theoretically or practically, and only an inconsistent physician will attempt to practice both. Osteopathy does not need to be bolstered up by the use of any therapeutical knowledge to be learned at any medical school. Each state association should adopt such rules as will require the resignation of all two-faced practitioners and prevent them from being taken in hereafter. Osteopaths cannot afford to compromise their position in regard to drug medication and should bar from their association all mixers and their ilk, who honor neither the profession of osteopathy nor medicine.[24]

In 1905 such an event as he called for took place when the Illinois Osteopathic Association asked one of its members, W. A. Hinckle, DO, MD, to resign. In an eloquent reply he wrote:

> Being a physician and not a sectarian practitioner I am heir to and privileged to make use of any and all therapeutic measures which the accumulated knowledge of centuries has shown to be of value, or which future learning may place within my reach regardless of its source or character. . . . Every physician must decide from his own experience and from the experience of others as to the relative value of the curative measures at hand and on the breadth of his learning, the accuracy of his judgment and his freedom to choose will decide his stature as a physician. This freedom your president informs me is neither desired nor

permitted in your society, I am given to understand that you prefer to share fellowship only with those who choose limitations rather than freedom. As membership in your society can therefore be purchased only at the price of intellectual liberty, I hereby present you with my resignation preferring rather the glorious isolation of unfettered thoughts and activities than the company of those who are slaves to creed and dogma.[25]

In spite of this organizational push to impose therapeutic purity by some state societies, more DOs were beginning to question the wisdom of rejecting all chemical and biological tools. With statistics gathered from clinical research demonstrating the efficacy of these proscribed agencies, the rationale upon which an osteopath could shun them became far more difficult to defend.[26] Some practitioners were forced to reconsider the issue when they themselves encountered failure in treating patients with manipulation alone. In 1908 Frank Furry, DO, MD, then vice-president of the AOA, told his colleagues of his own dilemma in a poignant account of caring for his daughter, who had contracted diphtheria.

> I had kept myself reasonably well posted on the serum therapy . . . and was opposed to the use of antitoxin. I chose osteopathy straight and we fought it out on that line and lost. No internal medication was used, excepting a hypodermic injection of strychnine to support the heart during the intubation process at the last. . . . The specialist who performed the intubation . . . called me a criminal in the presence of my dying child because I had not used antitoxin (which he claimed to be an absolute specific) and since thinking the matter over I do not know but that every member of our profession is a criminal just to the extent that he has failed to assist in the solution of this awful problem.[27]

From the beginning of the twentieth century students in osteopathic colleges were at least being exposed to information about biological and chemical agents, through their courses in toxicology, surgery, obstetrics, and practice, since instructors for the most part used the same textbooks employed in orthodox medical schools. Though many osteopathy teachers ignored or attacked the sections of such works dealing with the supposed benefits of these agents, except for anesthetics, antiseptics, and antidotes, other faculty members appeared less dogmatic. In 1906 Charles Teall, DO, who had succeeded Booth as the AOA inspector of schools, complained that too many medical notions were finding their way into the lectures. At Chicago, "'Broad Osteopathy' a science embracing everything was talked by the president of the senior class." In Boston, "elaborate detail in antiseptic, medicated douches, lotions, etc. is given, while the osteopathic part is taken for granted with 'of course find the lesion and remove.'" At Los Angeles, "certain formulas were on the board and copied by the students which will land them in jail, or at least give them trouble if used in most any state of the union for it was strict medical practice."[28]

In defending their schools, a number of faculty and administrators argued that there should be some classroom discussion of chemical and biological agents so that students could intelligently decide the merits of their use for themselves. Furthermore, it appeared to them that even more instruction in these modalities would have to be given, whether they liked it or not, if their graduates were to secure greater legal privileges insofar as surgery and obstetrics were concerned. In Illinois, for example, the medical act allowed for the granting of two types of licenses, one for a physician and surgeon and the other for a drugless practitioner. To be eligible for the first, candidates had to have graduated from medical schools approved by the state board of health. Such approval required the inclusion of a complete course in *materia medica*. Supporters of the Chicago school tried to change the law, but repeatedly failed. Consequently, in 1909 the college attempted to comply by adding "osteopathic *materia medica*" to the curriculum. It then applied for recognition but was turned down on the grounds that the subject was not adequately taught. The Littlejohns sued the board but eventually lost their case, whereupon ownership of the institution was transferred and the course was dropped.[29]

A similar situation occurred in California with a different outcome. Since 1906 DOs in California had been able to secure full physician and surgeon certification if they passed the same test required of MDs. However, in 1913 the law was amended to stipulate that anyone wishing to take the examination had to be a graduate of a college giving a minimum number of hours in specified subjects, including pharmacology. The Los Angles School therefore made the necessary changes and thus became approved by the composite California Medical Board.[30]

While the Chicago and Los Angeles colleges represented extreme cases, other schools were also expanding their curricula. Both Philadelphia and Boston offered optional courses on *materia medica* which did not appear in their annual catalogs.[31] Des Moines introduced a series of lectures called "Comparative Therapeutics" which it defended on the grounds that "in this way the osteopath will be better able to explain the practice of osteopathy to the minds of a public used to drugs."[32] Even Kirksville, after the founder retired from active control, began moving into previously prohibited areas. In 1911 its catalog description of the course in bacteriology noted, "vaccines, antitoxins and serum therapy with the values and ill effects resulting from the careless and improper use of each in practice are specifically and logically taught."[33]

These straightforward and roundabout efforts at integrating chemical and biological agents into the curricula were naturally opposed by the lesionists, a number of whom held important positions within the AOA hierarchy. At first they tried to cajole the colleges into withdrawing these subjects, but when this seemed a waste of time they decided to follow a more drastic course. At the 1914 convention in Philadelphia the Board of Trustees ruled that after 1916 "engaging in the teaching of drug therapeutics by any member of this association shall be cause for depriving of membership in this organization; and that participants in such training by the college shall be cause for refusal by the Association for recognition of such colleges as a cooperating institution."[34] Several months later the AOA board supported the successful lobbying efforts of a group of Oregon DOs who secured an amendment to their existing law which stated, "No school of osteopathy whose curriculum includes a course in materia medica, pharmacology or prescription writing is to be considered for the purpose of this act to be a regularly conducted school of osteopathy."[35]

Protest within the ranks nationally as well as in Oregon soon followed. Dr. Henry Bunting, publisher of the journal *Osteopathic Physician*, listed a number of orthodox remedies and hypothetical situations for his colleagues to consider:

> If you had an elderly patient whose body was eaten out with malignant cancer, dying by inches, would you yield to her entreaties and give her morphine? If you had a son who was a cretin would you give him thyroid extract? If your child had diphtheria would you use antitoxin? If bitten by a mad dog would you yourself take the Pasteur treatment? If you had a patient bleeding to death would you blanche the wound with adrenaline? . . . Would you use pumpkin seed to expel a tapeworm? Would you give an anemic organized iron? If you had a syphilitic patient would you use mercury or salvarsan or anything else now used to help that condition? . . . If you had a patient whose heart beat about 160 and you weren't sure the pulse was strong enough to count would you ever wonder if digitalis might not be a help in that one case?[36]

With Bunting for the first time publicly declaring himself to be in favor of teaching the use of biological and chemical agencies in the colleges, other DOs who had previously kept silent on this issue voiced their support. They were joined by those who, while not in favor of a separate course in *materia medica*, were nonetheless opposed to the board's ruling and the Oregon law on the grounds that neither the AOA nor the state legislatures had any business interfering with the colleges' right to determine their educational policy.[37]

With opinion steadily mounting in a direction favoring the teaching of *materia medica*, the founder, now eighty-seven years old, made an open appeal to his followers just before the start of the 1915 convention in Portland, where this matter was sure to be raised. Still warned: "There is an alarm at the door of all osteopathic schools. The enemy has broken through the picket. Shall we permit the osteopathic profession to be enslaved to the medical truth? As the father of osteopathy, I am making an international call for the Simon-pure DOs who are willing to go on the fighting line without being drafted for service."[38] Still's plea was unsuccessful. Bowing to pressure from its critics, the Board of Trustees revoked the previous year's directive, thus in effect both disavowing itself from the Oregon law and leaving the colleges free to teach what they wanted.[39] Slowly, the profession was coming out from under Still's shadow.

Although this action seemed to signal the dawn of a new era for the movement, a rather extraordinary chain of events occurred soon afterwards which temporarily restored the lesionists to power. During 1918 and 1919 some 650,000 persons in the United States and approximately 40,000,000 worldwide died as a result of what was known as the "swine flu," a particularly lethal strain of influenza virus which had surfaced after several decades of dormancy.[40] No specific vaccine or serum had been developed, nor was there any effective drug therapy that could shorten or minimize the course of the disease. In their treatment of the afflicted, most American physicians proceeded cautiously, isolating the patient, establishing satisfactory hygienic conditions, and carefully regulating fluid intake. Drugs were used only to relieve symptoms, as in the more common forms of influenza. Those MDs relying on Osler's textbook, for example, gave a dose of calomel during the day to open the bowels, 10 grains of Dover's powders at night to relieve the aches and pains, aspirin to reduce the fever, and strychnine in full doses in cases manifesting great cardiac weakness.[41]

Most DOs, on the other hand, while generally following orthodox procedures with respect to isolation, hygiene, and fluid intake, rejected drugs altogether, substituting manipulative measures in their place. Such therapy directed at the spine and rib cage would, according to its advocates, help normalize visceral functions and specifically build up resistance to and disperse fluid in pneumonia, which was a common sequela.[42] Many DOs in the field at the beginning of the pandemic reported to their journals how well they seemed to be doing in comparison to local MDs. With the profession then seeking additional evidence with which to pressure Congress to allow its members to serve in the military medical corps, a campaign to gather and publish statistics was launched and given wide publicity in the *Osteopathic Physician* and *JAOA*.[43] Between October 1918 and June 1919, a total of 2,445 DOs mailed in a summary of their results and a general description of their approach. Of 11,120 influenza cases treated by DOs during this period, there were but 257 deaths listed (a .2 percent mortality). Of 6,258 pneumonia reports, there were only 635 fatalities (a 10.1 percent mortality). These figures were compared to an estimated 12 to 15 percent influenza case mortality rate and 25 percent pneumonia case mortality rate of patients under the care of orthodox physicians.[44] Although the adequacy of their data collection methods and conclusions was laid open to serious question by MDs, the DOs were convinced that they had documented their therapeutic superiority.[45]

The impact of this experience upon the members of the osteopathic profession was quite significant; many of those who had doubted the applicability of manipulative therapy in the management of acute infectious diseases began treating such cases with their hands. Furthermore, as a result of surviving the flu, or knowing someone who had, patients who had previously patronized the DO only for joint and muscle disturbances, as well as individuals who had never frequented the office of an osteopath, now decided that they would rely on the DO as their family doctor. The osteopathic profession's belief in a wide applicability of manipulative therapy was again on the rise.

With sentiments towards biological and chemical agents diminishing, those DOs who supported the 1914 resolution and Oregon amendment reasserted themselves. In 1920 the Board of Trustees and the House of Delegates passed The Profession's Policy, which attempted to set definite restrictions on the DOs scope of practice. One section of this document embodied a standard college curriculum covering what it called "all the subjects necessary to educate a thoroughly competent general osteopathic practitioner." Neither pharmacology nor *materia medica* was listed. Training in certain types of drugs, including germicides and parasiticides, was given, but no mention was made of other agents, such as digitalis, colchicine, or vaccines, serums, and endocrines. A second section concerning legislation called for a revision of the model bill, incorporating language in each state law to allow licensees to use only those drugs "taught in the standard college curriculum which means the standard curriculum of the AOA" Rather than oppose these new proscriptions, school officials, seeking compromise, cooperated in their formulation.[46] The renaissance in osteopathic fundamentalism had influenced many of them as it had practitioners in the field.

Nevertheless, once the initial wave of renewed enthusiasm had passed, dissatisfaction with the new AOA policy became evident. The broad osteopaths, ending a discreet period of silence, again took the offensive, blasting away at what they felt was the intellectual vacuousness of the AOA position. However, many practitioners seemed more upset by the adverse effect the policy had on their efforts to obtain favorable laws. A majority of state legislatures continued to reject attempts to expand the legal scope of osteopathic practice vis-à-vis surgery and obstetrics as well as those drugs the AOA sanctioned. Many legislators refused to budge from their long-held view that before they would seriously consider their requests, the DOs would have to demonstrate that they received the same breadth of undergraduate training as did the MDs. This meant teaching the use of all generally recognized preventive and therapeutic measures.[47] Rather than blame the legislators, these DOs turned their wrath on the AOA leadership, arguing that in its stubborn insistence upon a limited instruction in biological and chemical modalities it was biting its nose off to spite its face.

In 1924, due to strong student pressure, administrators of the Chicago College announced that it would once again attempt to meet legislative demands by adding a comprehensive course in *materia medica* to the curriculum. E. S. Comstock, DO, secretary of the school, declared, "If we have sufficient faith in the osteopathic concept and in osteopathic principles, if they are sufficiently convincing to the logic and intelligence of the average human being, why should we fear the knowledge of drug action, when so often the untoward results outnumber the beneficial effects." The AOA board, however, was unimpressed, voting seventeen to one against such a course in osteopathic colleges, thus causing the Chicago school, threatened with a loss of its accreditation, to back down.[48]

The board's decision in this case helped to fuel the opposition. With increasing numbers of DOs in the field resenting the association's placing a restriction on their scope of practice, and with some of the schools seemingly on the verge of openly defying the provisions of the standard curriculum, the AOA leadership began to realize that a reconsideration of the issue was necessary. In July 1927, members of the board met with representatives of the Associated Colleges of Osteopathy and hammered out another compromise. That fall each school, with the board's blessing, would begin teaching a course called Comparative Therapeutics. What this would include was not made explicit; nevertheless, it was thought that the title would satisfy the lawmakers.[49] Initial reaction, however, proved otherwise, with some in government characterizing it as a mere subterfuge.[50]

Frustration within the ranks mounted. Scathing letters from prominent DOs against the amended policy filled the pages of osteopathic publications, while a few state societies formally demanded that the board immediately make the necessary changes.[51] Given this steady bombardment of criticism, the AOA Board of Trustees met with the college officials once more in the summer of 1929. This time they agreed to an outline of a course called Supplementary Therapeutics, which specifically mandated complete training in the use of biological and chemical agents. This proposal was then submitted to the AOA House of Delegates, which had the final say. It decided to make sure that the legislatures knew what the phrase "supplementary therapeutics" meant by adding pharmacology as one of its subheadings.[52] As a few of the more conservative colleges felt that adding pharmacology per se was going too far, the house the next year made teaching the subject "permissible" rather than "required."[53] Nevertheless, the significance of the 1929 resolution remained undiminished. The official policy of the AOA was now in favor of a truly complete and unlimited scope of practice and would not be reversed.

THE PUSH FOR HIGHER STANDARDS

W ith DOs increasingly duplicating the role and services of MDs the focus of the debate over the relative merits of osteopathy gradually shifted from its underlying philosophical and therapeutical beliefs to an analysis of its educational system. The central question became whether the standards maintained by osteopathic colleges were adequate to ensure the production of qualified physicians and surgeons.

THE REVOLUTION IN MEDICAL EDUCATION

The issue of standards did not apply solely to osteopathy. At the turn of the century medical education in the United States was noted for its disparities. On one end of the continuum was a relatively small number of prestigious university-affiliated colleges, on the other were the profit-motivated proprietary schools. Despite the gulf between the two types of institutions in terms of staffing, facilities, and equipment, licensing laws made it as easy for the graduates of one type to obtain the right to practice as it did the graduates of the other. Most existing boards of registration and examination either did not have the power to set meaningful standards for the colleges or had declined to do so.[1]

In 1904 the recently reorganized AMA formed its Council on Medical Education to suggest methods of improving academic requirements and to serve as an ongoing agency for advancing the association's policies. In order to determine the actual situation in the colleges, the AMA Board of Trustees the following year authorized the council to undertake a complete on-site survey and to rate all 160 MD-granting schools. Although the grading was reportedly lenient, only 82 were given Class A, or approved, ranking; 46 were placed on Class B, or probation, and 32 were designated as Class C, or unapproved. While this information was not revealed to the public, it was made available to each state licensing board for its consideration; as a result several of the boards decided they would henceforth refuse to examine graduates of schools not receiving the council's approval. Many colleges were thus motivated to begin making needed improvements, and others simply shut their doors. Between 1906 and 1910, the number of MD-granting institutions decreased by 29.[2]

The lay public's first detailed knowledge of the still generally lamentable school conditions came with the publication of Abraham Flexner's *Medical Education in the United States and Canada* (1910), an on-site survey carried out under the auspices of the Carnegie Foundation for the Advancement of Teaching and in cooperation with the AMA council. Though Flexner found some colleges upholding what he considered to be satisfactory standards, these constituted a decided minority. With respect to matriculation, only one of four was insisting upon either a high school diploma or liberal arts college credit as the minimum prerequisite for admission; the remainder were permitting even the barely literate to enroll. Most schools lacked fully equipped laboratories for the first two years of instruction, and in the third and fourth years too many students were not being given the necessary hospital and dispensary experience to prepare them for practice.[3]

In his report Flexner suggested several reforms. First, he urged that all proprietary schools be closed down. Since the United States then had far more MDs per 100,000 people than in the industrialized European nations, it was unlikely that the loss of these institutions and their graduates would lead to a physician shortage. He further recommended that each surviving college become an integral component of a major university, thus ensuring higher academic standards. Finally, he strongly suggested that the financing of medical education be altered. Since tuition fees could cover only a fraction of the expenses necessary to support an adequate program, other sources of income had to be cultivated.[4]

This survey had a considerable impact upon the American consciousness. In the era of muckraking journalism, Flexner's overall findings and vivid descriptions of individual schools made excellent copy and were widely circulated by the nation's press. Now in a position to mobilize public opinion, the various groups committed to change went forward in their efforts to accelerate the progress already being made.[5] In the twenty-five years following the appearance of the Carnegie Foundation study several significant improvements were made along the lines Flexner had laid down. First, the number of schools steadily dropped. Commercial and otherwise weak institutions were forced out of business as more state boards accepted the continually updated ratings of the Council on Medical Education.[6] By 1935 there were but 66 AMA-accredited colleges, 57 of which were connected with a university.[7]

Higher entrance standards were also set and maintained. In 1918 the council ruled that all incoming students had to have completed two years of college work. As of 1936, 83 percent of all matriculants exceeded this minimum, while 49 percent enrolled with a baccalaureate degree. The educational program itself was greatly enhanced, this due in large part to the changes in the colleges' fiscal condition. During the 1934–35 and 1935–36 academic years, 55 percent of the total income of all medical schools was raised through taxes, public and private general university funds, and philanthropy.[8] With the additional revenue these sources brought, the colleges built more completely outfitted laboratories, hired full-time basic science instructors (mostly PhDs), and upgraded their hospital and dispensary facilities.

These advances helped spur considerable progress on the postdoctoral level. With the schools' rise in quality, graduate programs ceased being "undergraduate repair shops."[9] In 1912 the AMA council conducted its first inspection of hospitals offering internships. From then through the mid-1930s, standards for internships were regularly strengthened as this additional year of training became all but obligatory.[10] In 1927 the council published its first list of approved residencies, and in 1933 the AMA established the organizational machinery to create boards of certification in the various specialties. These changes, along with those on the predoctoral level, would provide the American people with a more uniform corps of highly skilled MDs.

OSTEOPATHIC EVOLUTION

Flexner included all eight osteopathic colleges in his grand tour of the nation's medical schools. The impetus behind the inclusion of osteopathic medical education in Flexner's study probably came from Henry Pritchett, head of the Carnegie Foundation. He and Flexner attended a meeting in December 1908 with the members of the AMA Council on Medical Education to discuss the impending survey. According to the minutes:

President Pritchett related his experiences with an osteopath in a small Colorado town, making his lame leg an excuse for calling as a patient but seeking information about the osteopath's methods and what kinds of cases he treated. It was found he was treating even adenoids and appendicitis. Therefore, it was clear that osteopaths (at least this one) were diagnosing the same diseases which physicians were called upon to treat, therefore osteopaths should have the same training in fundamentals.[11]

While members of the council asked Pritchett about his encounter, they expressed no apparent interest in including osteopathic education in the Carnegie Foundation study. In his report, Flexner echoed and extended Pritchett's reasons for evaluating DO colleges as medical schools in his survey in spite of the differences in approach between DOs and MDs. Flexner argued:

> Whatever his notions on the subject of treatment, the osteopath needs to be trained to recognize disease and to differentiate one disease from another quite as carefully as any other medical practitioner. . . . Whether they use drugs or do not use them, whether some use them while others do not does not affect this fundamental question. . . . All physicians summoned to see the sick are confronted with precisely the same crisis: a body out of order. No matter what remedial measures they include—medical, surgical, manipulative—they must ascertain what is the trouble. There is only one way to do that. The osteopaths admit it when they teach physiology, pathology, chemistry, microscopy.[12]

Having, for the purpose of his analysis, placed the osteopathic profession on an equal footing with orthodox medicine, Flexner was quick to emphasize that not one "of the eight osteopathic schools is in a position to give such training as osteopathy demands." The teaching of anatomy, for example, was "fatally defective." Most of the students' time during this course was spent listening to lectures; too few cadavers were available to provide adequate laboratory dissection. This pattern characterized the other basic sciences as well. "A small chemical laboratory is occasionally seen," Flexner noted.

> At Philadelphia it happens to be in a dark cellar. At Kirksville, a fair sized room is devoted to pathology and bacteriology; the huge classes are divided into bands of 32, each of which gets a six weeks course following the directions of a rigid syllabus, under a teacher who is himself a student. . . . A professor at the

Kansas City school [the Central College] said of his own institution that it had practically no laboratories at all; the Still College at Des Moines has in place of laboratories laboratory signs; the Littlejohn at Chicago, whose catalog avers that the "physician should be imbued with a knowledge of the healing arts in its widest fields, and here is the opportunity" has lately in rebuilding wrecked all its laboratories but that of chemistry without in the least interfering with its usual pedagogic routine.

Clinical instruction fared no better. "The osteopath," he declared, "cannot learn his technique and when it is applicable except through experience with ailing individuals. And these for the most part he begins to see only . . . after receiving his DO degree." Bedside training was, in fact, either very limited or nonexistent. The Kirksville College had the largest hospital, a mere fifty-four beds, while the Chicago school had twenty, the Pacific College fifteen, the Boston school ten, and the Philadelphia college three. The Des Moines, Kansas City, and Los Angeles schools had none at the time of his visit. Outpatient contact was similarly restricted. Each of the colleges operated a pay clinic that was staffed by the faculty, in which student participation seems to have been limited to the care of charity cases.[13]

In characterizing the entire osteopathic educational program, Flexner wrote:

> The eight osteopathic schools now enroll over 1,300 students who pay some $200,000 annually in fees. The instruction furnished for this sum is inexpensive and worthless. Not a single full time teacher is found in any of them. The fees find their way directly into the pockets of the school owners, or into school buildings, and infirmaries that are equally their property. No effort is anywhere made to utilize prosperity as a means of defining an entrance standard or developing the "science." Granting all that its champions claim, osteopathy is still in its incipiency. If sincere its votaries would be engaged in critically building it up. They are doing nothing of the kind.[14]

Angry protests by school officials and other DOs greeted the publication of Flexner's report. Responding to his critique, the AOA Board of Trustees declared, "We have no apologies to offer for our colleges. They have done well, and we take pride in their attainments and in their ambitions and determinations to teach most thoroughly and scientifically all that pertains to disease in all its phases and manifestations. We demand that they be allowed to do this, according to the needs of our profession and not in accord with the wishes of any self-appointed, self-seeking, tyrannical and prejudicial judges."[15] Interestingly, this view was not completely shared by the AOA Committee on Education. In its annual report for 1910, it substantially agreed with Flexner on the problems of low entrance standards, poor basic science laboratories, lack of sufficient clinical facilities, and an inadequate teaching corps. Its own surveys had noted the same deficiencies, albeit in less caustic language.[16] However, unlike Flexner, who evaluated the schools with an ideal in mind, DO inspectors considered themselves pragmatic to the extent that they recognized the limited possibilities for amelioration under existing conditions. Reform, they believed, would have to be slow.

The twenty-five years following the issuance of the Flexner report saw some improvements in college requirements and in the quality of training offered; nevertheless, osteopathic institutions did not keep pace with the changes incorporated by the MDs. With respect to preprofessional education, the AOA Board of Trustees in 1920 stipulated that henceforth each school must maintain an entrance standard of no less than a high school diploma or its equivalent to keep its accreditation rating. However, no attempt was immediately made to enforce this provision, and it was not until the early 1930s that all the schools appeared to be fully complying.[17] Those in favor of further stringency in entrance requirements were a decided minority. The Los Angeles college established a compulsory one year of preprofessional qualification in 1920, but this was in response to a new California law. Most DOs sided with Dr. George Laughlin, who in 1925 observed, "We make a mistake as a profession when we attempt to ape the medical man in matters of requirements."[18] Laughlin, the founder's son-in-law and then head of the Kirksville College, argued that the requirement by MD institutions of two years of prior college work was hurting the underprivileged, since they could least afford the cost of additional schooling. As many of these disappointed students came from farms and small towns, the standard had the indirect effect of causing a decline in the percentage of recent MD graduates deciding to locate in sparsely populated areas. Without this qualification, DO schools could meet the needs of the economically disadvantaged student and help alleviate a growing rural physician shortage.[19]

Whatever the merits of Laughlin's views, the main reason militating against a further increase in preprofessional requirements was the economic condition of the colleges themselves. Although all of the schools had evolved into nonprofit institutions, the sources of their revenues remained the same. They received no direct tax support, no general university funds, and, in comparison with MD institutions, little outside philanthropy. In 1932 reportedly 92 percent of the gross receipts of all the colleges was secured from tuition fees alone.[20] Given this form of financing, the schools' very survival depended on their ability to obtain a base line of new matriculants each fall. If they set the preprofessional entrance standard at the MDs' level of two years or more, the osteopathic schools would drastically cut their pool of eligible applicants, and the number of students necessary to meet expenses would very likely not be reached.[21]

During this era a large percentage of the schools' annual tuition income was devoted to establishing more permanent facilities. In 1921 the Los Angeles College moved to a new campus, where three large buildings were erected over the next decade. The Kansas City College of Osteopathy and Surgery, founded in 1916, had two homes before finding a suitable location four years later, where it raised five new structures by 1933.[22] The Chicago school left the downtown area for the Hyde Park section of the city in 1918, renovating a large four-story working girls' residence to provide classrooms, laboratory, hospital, and clinic space. The Des Moines College in 1927 relocated from one entire office building to another, while the Kirksville College of Osteopathy and Surgery added a new facility for laboratories and classrooms along with a second hospital. Only the Philadelphia school, which in 1929 established a new campus, costing $1.1 million, was able to finance its plans through private donations.[23]

The educational program of the schools underwent a number of important changes between 1910 and 1935. The colleges added a mandatory fourth year, introduced a graded curriculum, and integrated the teaching of biological and chemical agents into the course of study. As a result, in its promotional literature the profession could boast that in terms of subjects presented and time devoted to them, MD and DO schools were equivalent. Indeed, on paper osteopathic institutions offered students a few hundred more hours of training than the typical orthodox medical college. However, this was a deceptive figure. Although the length of basic science courses in osteopathic colleges was greatly expanded, the instruction itself continued to be weak.[24] By the early 1930s some preclinical teachers were employed on a full time basis, but few of these DOs possessed a graduate degree in the subjects they taught. The new buildings provided more adequate facilities, yet the equipment remained meager, and most laboratories were fitted out with the barest of necessities.

Money that could have purchased additional, improved apparatus had to be diverted into mortgage payments. Finally, the courses were often not as encompassing as those in allopathic colleges, partly because the pre-professional backgrounds of MD and DO matriculants differed. Osteopathic curricula, for example, included elementary biology and chemistry, which medical students had mastered before beginning their formal professional education.

Clinical training was also beset by severe difficulties. All of the schools were operating larger hospitals in the mid-1930s than previously; however, most were still quite small. While most MD colleges easily surpassed the minimum of 200 beds available for teaching purposes in the guidelines set by the AMA Council on Medical Education,[25] the Chicago, Des Moines, Kansas City, Kirksville, and Philadelphia osteopathic schools averaged only 66 beds apiece.[26] Where a minimum of 2,000 curriculum hours were devoted to bedside and outpatient teaching at MD-granting institutions, an average of approximately 700 hours of training were provided by these five colleges.[27]

The one osteopathic school that was able to offer clinical training approaching that found in orthodox medical schools was the College of Osteopathic Physicians and Surgeons (COP&S) of Los Angeles. This was made possible through its utilization of a 203-bed public hospital, which enabled each student to receive 1,770 hours of inpatient and dispensary experience.[28] The establishment of this institution was an unintended by-product of the "standardization of hospitals" plan inaugurated in 1918 by the American College of Surgeons, which eventually partnered with the AMA in running this program. These groups required that any hospital seeking their approval for the purpose of training graduate MD physicians prohibit DOs from having admitting or staff privileges. As a result, osteopaths throughout the country who had managed to secure such rights found them abruptly terminated.[29] The Los Angeles County government, responding to pressure from DOs, who were numerous in the metropolitan area, built and opened a separate public hospital for the training and service needs of COP&S, whose faculty had been denied privileges at the existing public facility.[30] Unfortunately for the profession, this type of arrangement was not repeated elsewhere.

With clinical experience in the colleges generally limited, it is hardly surprising that postgraduate training was also far from satisfactory. By the middle 1930s there were no more than eighty osteopathic hospitals in the country, of which only one-quarter were offering opportunities for advanced work.[31] In 1932 there were but seventy-five internships, few approaching the standards governing MD programs.[32] Formal residencies were scarcer. Those osteopaths who could not enter one of these had to learn their specialty by attending short courses given at the colleges or by taking a preceptorship with a private practitioner. Clearly, such conditions, as well as those on the undergraduate level, left much to be desired.

THE PRICE OF LOWER STANDARDS

Although osteopathic colleges during most of this period attempted to prepare their students to become full-fledged physicians and surgeons, their graduates faced difficult problems in being licensed as such. By 1937 only twenty-six legislatures had agreed to extend them privileges commensurate to those enjoyed by the MDs, and in some of these states a majority of DOs continued to be ineligible, since sixteen mandated pre-professional college work and eight stipulated a year-long internship.[33] Furthermore, even when these requirements were met, other hurdles remained. In those jurisdictions where DOs had to be examined before medical or composite boards, they fared rather poorly on the same written tests taken by allopathic candidates. Between 1927 and 1931, for example, only 48 percent passed compared to 95 percent of the MDs.[34] Consequently, many DOs avoided these examinations altogether, choosing an unlimited-license state whose tests were devised and graded by an osteopathic board and where the rate of failure was negligible. This reinforced the disproportionate geographical distribution of DOs that had existed since early in the century and that had been directly related to the location of the colleges.[35]

Unable to convince state legislatures to eliminate independent osteopathic boards, the MDs adopted the strategy of lobbying for a common test in the basic sciences that was to be taken prior to an actual licensing examination. This preliminary exam, which would be required of MDs, DOs, and chiropractors, would cover such subjects as anatomy, physiology, bacteriology, and pathology and would be written and administered by a separate committee independent of the licensing boards. In 1925 Connecticut and Wisconsin became the first to create basic science boards; they were followed by Minnesota, Nebraska, and Washington two years later.[36] In opposing such measures, the AOA House of Delegates argued, "Such an arrangement creates superfluous and unnecessary machinery of administration, erects another financial barrier to the recent graduate who is starting upon his life's work of helping the suffering; is an inadequate practical test in the fundamental subjects considering the varying viewpoints and methods of the different schools of practice; eliminates reciprocity between existing osteopathic boards which are now functioning in a manner to insure the public osteopathic physicians who are well qualified; and furnishes the opportunity for domination by so-called 'regular medicine.'"[37] The real fear of the osteopathic profession, however, was that their graduates would not be able to do as well as the allopathic practitioners, and this was soon confirmed by early results. In 1930, before seven basic science boards, the pass rate was 88.3 percent for MDs, 54.5 percent for DOs, and 21.9 percent for chiropractors.[38] As a consequence, osteopaths began avoiding states that mandated such exams. One AOA spokesman noted, "In the three states of Minnesota, Nebraska, and Washington where the figures are available to make such a comparison, we find that those states jointly licensed 158 DOs in the two and one half year period prior to the adoption of the basic science boards. In the two and one half year period since . . . they have licensed but 35 or about one-fifth as many."[39]

In accounting for their mediocre performance on basic science as well as medical board tests, the DOs asserted that they were being discriminated against, since osteopathic emphases were being ignored. If such examinations contained a fair number of questions bearing upon the mechanics of vertebral articulations or upon the role of nerves in controlling physiological functions, they argued, the results would be quite different.[40] This claim may have had some validity; however, it seems unlikely that these alleged biases contributed significantly to the rate of failure by DOs. A more likely reason is that the MDs as a group had a superior overall educational background.

THE GREAT LEAP FORWARD

With almost one-half of the states refusing to grant DOs unlimited privileges, with an ever-increasing number of states setting preprofessional and postdoctoral requirements that most DO graduates could not fulfill, and with DOs doing so poorly on outside examinations, the osteopathic schools recognized a need for fundamental change in the structure and quality of their educational programs. A mere continuation of their slow, evolutionary approach to reform was not likely to achieve the privileges their students sought, and it could conceivably cause DOs to lose what legal ground they had already gained.

The specter of this second possibility was raised by a 1934 survey of four osteopathic schools made by two Canadian academicians, Frederick Etherington, MD, and Stanley Ryerson, MD, as a consequence of a small number of DOs in Ontario seeking additional practice rights from the provincial government. Comparing these U.S. osteopathic institutions with the province's three medical colleges, Etherington and Ryerson showed that the DO schools were characterized by inferior laboratories and equipment, smaller hospital and clinic facilities, lower matriculation requirements, and less-qualified faculties. As osteopathic colleges did not, in their opinion, adequately prepare students to become physicians and surgeons, their graduates should not be licensed as such.[41] These findings and conclusions were widely publicized by the AMA, which brought this survey to the attention of United States lawmakers.[42] Put on the defensive, the DOs maintained that this "so-called inspection" was hastily done, that the examiners were obviously prejudiced, and that much of the information published was either misleading or blatantly untrue.[43]

While some state legislators gave the DOs the benefit of the doubt, others called for an unbiased legislative inquiry. With a few states on the brink of authorizing such investigations, the Associated Colleges of Osteopathy hired an outside consultant to prepare his own separate and confidential evaluation. The investigator, L. E. Blauch, PhD, was a nationally known educator who several years earlier had headed a Carnegie Foundation study of the curricula of American dental schools. In 1936 Blauch, with the approval and cooperation of the AOA, accompanied the chairman of the Bureau of Professional Education on his regular inspection of five osteopathic colleges. If the DOs anticipated a more favorable portrait of educational conditions, they were to be disappointed. In a detailed and dispassionate series of reports, Blauch cited the same deficiencies noted in the Canadian survey.[44] Obviously, should state legislatures decide to commission their own investigations, the legal status of the colleges would be placed in considerable jeopardy. In view of this prospect, educational reformers within the osteopathic profession now gained the upper hand.

The MDs had made their largest strides in raising standards during the first twenty-five years after the Flexner report; the DOs made theirs during the second. One of the earliest reforms they effected was in admission requirements. In 1934 the Philadelphia college began enforcing a prerequisite of one year of college, and in 1937 it followed COP&S, which twelve months earlier had increased its minimum to two years. The Chicago and Kansas City schools went directly from requiring a high school diploma to a prerequisite of two years of college, in 1938, while the Des Moines and Kirksville colleges, meeting an AOA-imposed deadline, instituted a one-year condition in 1938 and a two-year requirement in 1940.[45]

As anticipated, enrollment suffered. In 1937 there were 1,977 students in the six accredited colleges; by 1940 the number had dipped to 1,653, a decline of 21 percent. This trend was accelerated by the entry of the United States into the Second World War, which drastically reduced the number of undergraduate college students available to enter any professional school. In 1945 total osteopathic enrollment had shrunk to 556— by far its lowest point in the century.[46] Immediately after the war, the AOA hired a full-time vocational counselor, who visited liberal arts colleges across the country, meeting with placement officers and students and acquainting them with the osteopathic profession.[47] This campaign, in conjunction with the schools' individual recruiting drives, which were aimed not only at current undergraduates but at returning veterans, soon brought the desired results. In 1947 total matriculation had climbed back to where it had been prior to the establishment of the two-year prerequisite, and it remained stable for more than a decade. Indeed, during this period the ratio of qualified applicants to available freshmen positions rose to roughly two to one, making admission into the colleges competitive for the first time.[48] This served to strengthen the credentials of osteopathic students and encouraged each of the schools to raise its entrance requirements to three years of college work. Los Angeles did so in 1949; Chicago, Des Moines, and Kansas City in 1952; and Kirksville and Philadelphia in 1954. By 1960, 71 percent of all new osteopathic students were entering with a bachelor's or an advanced degree.[49]

As higher prerequisites for admission were being introduced, osteopathic schools were enriching their basic science curriculum. Although the total combined average number of hours in anatomy, physiology, biochemistry, pathology, and microbiology remained virtually unchanged from 1935–36 to 1948–49, the percentage of time spent in the laboratory as opposed to the lecture hall increased from 48 percent to 59 percent, a figure that continued to climb in subsequent years.[50] Three of the schools—Chicago, Kansas City, and Los Angeles—erected new basic science buildings, while other colleges upgraded existing facilities and equipment. Furthermore, after 1945 each of the schools hired more full-time instructors with MS and PhD degrees, thereby enhancing the quality of their faculties.[51] An even greater transformation occurred in the second half of the undergraduate osteopathic curriculum. Actual bedside and outpatient experience for each student was increased in all six schools from an average of 862 hours in academic year 1935–36, to 1,883 hours in 1948–49, to 2,214 in 1958–59.[52] This can be attributed both to the expansion of the colleges' hospitals, from a combined total of 530 beds and bassinets in 1935 to 1,334 in 1959, and to the fact that each of the schools made arrangements with other osteopathic hospitals for the training of externs.[53]

Quite a few of these changes in undergraduate education were possible only because the schools put themselves in more financially secure positions. Since the annual number of qualified applicants far exceeded the freshman places available, the colleges could institute sizable tuition boosts without jeopardizing the number of matriculants. Between 1935 and 1960 fees climbed from an average of $223 to $900 per year.[54] Outside sources were also solicited. In 1943 the AOA launched what became known as the Osteopathic Progress Fund. With student enrollment then dropping towards dangerously low levels and with several of the schools facing the prospect of having to close their doors, DOs in the field were pressured to contribute. By mid-1944, at the end of the first campaign, an impressive total of $962,535 had been subscribed and directly channeled into the college treasuries.[55] In 1946 a new, continuous, Osteopathic Progress Fund program was organized which raised $8,956,625 between then and 1961.[56] This era also marked the genesis of federal support. In 1951 the U.S. Public Health Service awarded all six schools renewable teaching grants previously designated only for MD and dental colleges. By 1956 this source of income amounted to $383,000 a year. Another federal program aiding the schools came in the form of hospital construction funds made possible under the Hill-Burton Act of 1946.[57] Among the major grants made under this law was one awarded to the Kansas City college for a new clinic, one to the Kirksville school for a modern inpatient facility, and a third to COP&S for a rehabilitation center.

The advances in predoctoral education during this period were accompanied by significant changes on the postgraduate level. In 1936 the AOA Bureau of Hospitals undertook its first inspection of institutions offering internships. Since the primary objective of the association was to provide a position for every new graduate, requirements were initially set low in order to qualify as many of their hospitals as possible.[58] During the Second World War the DOs, who as a group were exempt from the draft and had been declared ineligible for voluntary service with the military medical corps, began taking care of the clients of inducted MDs. With allopathic hospitals still refusing DOs admitting and staff privileges, satisfied new patients stepped forward to help underwrite the costs of building and maintaining separate private osteopathic institutions. In 1945 there were approximately 260 osteopathic hospitals operating in the country, more than triple the total of a decade earlier.[59] This increase in turn served to alleviate the internship shortage, and by 1951 available positions had surpassed the number of that year's graduating seniors, thus making possible a toughening of standards.[60] In 1947 the Bureau of Hospitals made its first inspection of osteopathic residency programs. That year, 71 were approved; by 1959 there were 389 available.[61] As formal residencies increased in number, the requirements governing them, as well as the process of certification of specialists (under machinery created by the AOA in 1939) were considerably strengthened.[62]

The push for higher standards between 1935 and 1960 resulted in progress on the legal front. At the end of this span of time, the number of states in which DOs became eligible for unlimited licensure rose to 38; osteopathic schools were now able to meet the requirements of certain medical boards and other governmental agencies which had been em-

Table 1. Osteopathic Specialty Boards Established Prior to 1960

Speciality	Date Incorporated
Radiology	1939
Surgery	1940
Opthalmology and Otolaryngology	1940
Pediatrics	1940
Proctology	1941
Neurology and Psychiatry	1941
Internal Medicine	1942
Pathology	1943
Dermatology	1943
Rehabilitation Medicine	1954
Anesthesiology	1956

Source: 1998 Yearbook and Directory of Osteopathic Physicians (Chicago: American Osteopathic Association, 1998), pp. 721–40.

powered to approve them; and DO graduates possessed a preprofessional background and postgraduate training matching or exceeding the minimum called for by each state.[63]

Osteopathic performance on outside examinations also showed significant gains. While from 1942–44 to 1951–53 results obtained by MDs and chiropractors on basic science tests remained virtually unchanged, the

Table 2. Basic Science Board Examination Results for MDs, DOs, and Chiropractors, 1942–44 through 1951–53

Period	MD Examinees			DO Examinees			Chiropractic Examinees		
	Examined	Passed	%	Examined	Passed	%	Examined	Passed	%
1942–44	6,339	5,442	85.5	545	285	52.2	59	23	38.9
1945–47	8,628	6,935	80.3	526	306	58.1	134	44	30.5
1948–50	8,921	7,768	87.0	1,032	629	60.9	1,489	524	35.1
1951–53	9,693	8,448	87.1	903	723	80.0	579	217	37.4

Source: "Medical Licensure Statistics," *Journal of the American Medical Association* 122 (1943): 11; 125 (1944): 143; 128 (1945) 123; 131 (1946): 133; 134 (1947): 283; 137 (1948): 638; 140 (1949): 321; 143 (1950): 471; 146 (1951): 372; 149 (1952): 479; 152 (1953): 450; 155 (1954): 482. Beginning in 1954, results were not reported by type of practioner.

Table 3. Examination Results before Medical and Composite Licensing Boards for U.S.-Trained MD and DO Physicians and Foreign Medical Graduates, 1940–44 through 1955–59

Period	U.S. MD Examinees			U.S. DO Examinees			Foreign Medical Graduates		
	Examined	Passed	%	Examined	Passed	%	Examined	Passed	%
1940–44	27,158	26,291	96.8	940	589	62.6	7,152	3,371	47.1
1945–49	26,840	26,005	96.8	881	618	70.1	2,943	1,339	45.4
1950–54	27,052	26,145	96.6	1,021	810	79.3	6,118	3,270	53.4
1955–59	30,184	28,903	95.7	1,174	954	81.2	11,192	6,787	60.6

Source: DO figures from 1940 to 1945 weere culled from "Report of the American Association of Osteopathic Examiners: 1952," microfilmed, American Osteopathic Association Archives, Chicago. Subsequent DO data and all MD information derived form "Medical Licensure Statistics," *Journal of the American Medical Association* 116 (1941): 2025; 119 (1942): 145; 122 (1943) 94; 125 (1944): 126; 128 (1945): 106; 131 (1946): 114; 134 (1947): 260; 137 (1948): 609; 140 (1949): 298; 143 (1950): 446; 146 (1951): 344; 149 (1952): 450; 152 (1953): 421; 155 (1954): 452; 158 (1955): 275; 161 (1956): 341; 164 (1957): 426; 167 (1958): 573; 170 (1959): 573; 173 (1960): 387.
Note: Osteopathic data include candidates from both AOA-accredited and nonaccredited schools. U.S. MD data reflect those graduated from AMA-accredited institutions only.

DOs went from a 52 percent to an 80 percent pass rate. Substantial increases were also made before state medical and composite boards of licensure. Here too, while the results of graduates from U.S. medical schools remained consistent from 1940–44 to 1955–59, the rate of passage for DOs climbed from 63 percent to 81 percent. Clearly, whatever educational problems remained, the DOs had placed their academic house upon a more solid foundation.

7

A QUESTION OF IDENTITY

Osteopathy as originally conceived by Andrew Still was a radically different approach to healing. Its philosophy, view of pathology, and system of patient care shared little with the components of orthodox medicine. Indeed, the founder cast himself and his followers as nothing less than revolutionaries seeking to overturn the entrenched allopathic order. However, as the DOs came to adopt a multidimensional conception of disease and as their scope broadened, the clarity of the objective differences between the two groups began to fade. This trend would later be accelerated by two further developments: first, the progress made in improving their educational system and opportunities for complete licensure, as just described; and second, by their growing reliance upon orthodox medical modalities. However, these transformations were not accompanied by a commensurate change in the general public's perception of who the osteopathic practitioner was and what he or she did. As a result, most DOs would suffer, to varying degrees, from status inconsistency.

THE DISPLACEMENT OF OSTEOPATHIC MANIPULATIVE TREATMENT

The one feature of osteopathic practice that most readily distinguished the DOs from the MDs was, of course, manipulative treatment. After 1930, however, the application of this modality in total patient management began a steady decline. This trend can be attributed to, first, institutional changes, that is, alterations in the social structure of the colleges, the hospitals, and office practice; and second, to scientific changes, that is, transformations in the DOs' knowledge base.

The improvements that were undertaken in the colleges beginning in the 1930s were all initiated with the idea of raising their graduates' chances of becoming eligible for and passing unlimited licensure examinations. Since the distinctive elements of osteopathic education had no specific relevance to these goals, the colleges had no incentive to emphasize or build up this area of the curriculum. Indeed, some of the improvements were often instituted at the expense of distinctive osteopathy. Many of the full-time non-DO teachers hired to upgrade the standards of basic science instruction, for example, did not have the background necessary to integrate osteopathic theory into their lectures, as their predecessors had done.[1] Also, the time spent on the subjects of pharmacology and surgery was increased to meet state requirements, and this seemed to have a detrimental effect on osteopathic instruction. The consequence, complained George Woodbury, DO, of Los Angeles, was that too many students were becoming "sadly confused and sorely disillusioned before their day of graduation.... The lavish display of therapeutic methods and modalities explained and utilized in college, hospital and clinic demonstrations had a tendency to weaken the emphasis on and minimize the need and value of distinctly osteopathic procedures."[2]

In the first osteopathic hospitals, the DO who admitted the patient would perform the surgery or at the very least handle the patient's pre- and postoperative manipulative care. However, with the advent of larger facilities and the establishment of a more clear-cut division of labor between DO specialists and general practitioners, the latter were less involved in such management. As a result, osteopathic manipulative treatment (OMT) waned. Besides claiming that they were "too busy" to administer such treatment themselves, DO surgeons increasingly viewed such intervention as impractical from a technical standpoint, particularly when the patient was in an oxygen tent, or hooked up to assorted monitors and intravenous lines. Indeed, many DO surgeons argued on these grounds alone that OMT was only suitable in ambulatory settings.[3]

This deemphasis and growing exclusion of OMT from the hospitals had a significant impact upon the colleges' postdoctoral programs. In his 1946 AOA presidential address, Dr. C. Robert Starks, referring to the undergraduate students, cited a fellow practitioner who had complained, "As soon as these individuals are graduated and put into the osteopathic hospitals they are immediately 'deosteopathized' [so] that by the time they finish their internships the osteopathic phase of their training has been discredited in large measure by osteopathic surgeons and other members of the staff, and those individuals go into practice with an apologetic attitude towards the osteopathic phase of their professional work."[4] Needless to say, graduates who went on past the internship to take hospital-based residencies were even more likely to develop a negative opinion of OMT.

After the war the AOA began to respond to this criticism. In 1947 its Committee on Hospitals announced that it would start enforcing its policy, in effect since the beginning of the AOA accreditation program, that house staff record distinctive osteopathic diagnostic and therapeutic procedures on all patient charts. The hospitals, however, offered little cooperation, and within five years the advocates of enforcement were labeling their own efforts a failure.[5]

Institutional changes at the level of office-based DOs were also affecting OMT use. With the shift from an essentially chronic to a broad-based practice, DOs saw more patients during the average workday. This increase in demand served to reduce the frequency and length of osteopathic treatment. To make more efficient use of their time, some practitioners turned to physical modalities that did not require their continued presence, such as the "spinalator," a device that could be set to manipulate vertebrae mechanically. However, to a far larger number of DOs, pharmacotherapy seemed a much more convenient substitute. The efficiency of writing a prescription or giving an injection over administering OMT appears to have been, in itself, a significant factor in a DO's decision on how he or she would manage a given patient.[6] Also critical in this regard were the patients' expectations of how they should be treated. Clients simply seeking physicians and not previously socialized into the traditional osteopathic approach were less likely to desire or expect OMT for disorders not directly involving the musculoskeletal system.[7]

As noted, scientific issues also had an impact on the relative frequency of OMT. Early basic research on animals, the bulk of which was carried out by Dr. Louisa Burns and her associates, lacked valid controls and provided at best only inferences that the osteopathic lesion played any direct or indirect role in the pathogenesis of visceral disease in humans. Indeed, some DOs came to feel that in many illnesses they treated the presence or absence of the lesion was irrelevant. In 1916, during the fight over the relative merits of the diphtheria antitoxin, Dr. Henry Bunting, editor of the journal *Osteopathic Physician*, asked his colleagues:

Would we rather hang on to our dogma that—no matter what the facts show— *it has always got to be a mechanical lesion?* Nothing is easier to prove in the case of diphtheria, at least, that the word "mechanical" has no business to be inserted as a necessary condition for getting that disease. The exciting cause is *vital*, not *mechanical*—the Klebs-Loeffler bacillus. Inject 100 guinea pigs, each of 250 grams weight, with an equal amount each (or 1–100th part) of minimum lethal dose of diphtheria toxin. Each guinea pig will be "sure dead" in 96 hours. Repeat the experiment with 1000 guinea pigs, the thousand will die. Repeat it with 1,000,000. The million will die on the same schedule. Does this mean anything? What caused the disease? What killed? Some unknown and different anatomical lesion in the case of each guinea pig, or the well known Klebs-Loeffler bacillus through its toxins.[8]

Carefully controlled research on the lesion under osteopathic auspices began in the late 1930s as a byproduct of the search by the profession for philanthropic support of its schools. J. Stedman Denslow, DO (1906–1982), then at the Chicago College, met with Alan Gregg, PhD, of the Rockefeller Foundation, who advised that outside funding might be more readily secured if the DOs scientifically demonstrated that they had a distinctive contribution to make to the healing arts. Through Gregg's help, Denslow, who had decided he would prepare himself for a research career, was introduced to a number of prominent neurophysiologists. Among those who provided him with counsel and assistance were Ralph Waldo Gerard, MD, PhD, of the University of Chicago, and Detlev Bronk, PhD, of the University of Pennsylvania, both leaders in the new field of electromyography.[9]

Denslow, after he had built one of the early differential amplifiers and recorders for simultaneous electromyographic observations of paravertebral musculature, moved to Kirksville to launch his research. Unlike Burns, he purposefully confined himself to asking limited, testable questions regarding the lesion phenomena, particularly what local neurophysiological manifestations were associated with those spinal areas designated as lesioned through palpation. Between 1941 and 1943 Denslow and his associates published four articles in two prominent nonosteopathic basic scientific journals, demonstrating that the motor neurons (anterior horn cells) at those segmental levels in the spinal cord associated with musculoskeletal stress had lower reflex thresholds than those at other, "normal," levels in the spinal cord. This was shown by applying measured amounts of pressure necessary to evoke contractions of the paravertebral muscles at that segmental level. These muscular contractions were detailed and recorded electromyographically. He further found that reflex motor

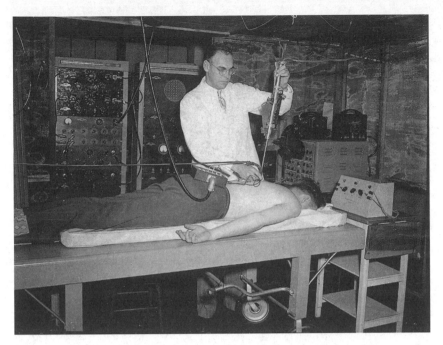

J. Stedman Denslow, DO, conducting electromyographic research
(early 1950s). *Courtesy of Still National Osteopathic Museum.*

thresholds of the paravertebral muscles, that is to say, their motor neurons, differed at different levels of the trunk, and that the reflex thresholds for both muscle contraction and pain were low in lesioned areas as
compared to nonlesioned or normal areas.[10]

After the Second World War Denslow was joined in his work by Irwin
M. Korr (b. 1910), who had received his PhD from the University of
Pennsylvania. Continuing this general line of experimentation, the two
men demonstrated in 1947 that diffuse and remote stimuli from many
sources preferentially excited the motor neurons of lesioned segments
while nonlesioned segments remained quiescent. Their research indicated
that the neurogenic mechanism responsible for this phenomenon was facilitation (that is, reduced threshold of excitability) of the motor neurons
of the spinal cord. Although the source of the facilitation was not at the
time conclusively demonstrated, Korr and Denslow hypothesized that it
might have its origin in a sustained afferent muscle bombardment from
segmentally related somatic or visceral structures. These impulses would
have the effect of lowering the threshold of excitability of the neurons at
the segmental level in the spinal cord's associated stress area.[11]

This research effort at Kirksville, which continued for decades, was of great import to the profession. It provided the first objective evidence of the presence of what the DOs had discerned through palpation and designated as the "osteopathic lesion"; it showed that a DO could do reputable studies on the lesion phenomenon and have the results accepted by the outside scientific community; and it paved the way for federal support of osteopathic research. However, the investigations, while breaking new ground, could not resolve two key questions. First, what was the significance of the lesion, or what was later called "somatic dysfunction," in the etiology of disease? Second, what effect, if any, would the elimination of the lesion through manipulation have on the disease process?

Controlled clinical research on manipulation might have been able to resolve this second issue; however, no such studies were undertaken during this era. The failure to pursue this course can be attributed in large part to the serious methodological challenges inherent in such a project. Though one could easily standardize the content and strength of a pill, it would be most difficult to have the same control over the amplitude and velocity of physical manipulation. Furthermore, with pharmaceuticals it was relatively simple to set up a single- or double-blind study with a capsule. Neither the patient nor the doctor would be able to distinguish the test drug from the placebo. However, what would constitute a manipulative placebo? The fact that one could not easily eliminate the subjective element from clinical studies on manipulation convinced those DOs within the AOA who controlled the association's limited funding for scientific projects to put their energy and money into basic research.[12] The clinical investigations that were carried out and published in osteopathic journals consisted of a small number of case studies, many of which were anecdotal in content.

Thus, DOs were told by their colleagues that the lesion was significant and that OMT worked, but they had to accept the concepts on faith or on circumstantial evidence. A growing number could not. As Louis Chandler, DO, of Los Angeles, noted in 1950, "Too much still seems to be in the realm of uncertainty both as to what will result from manipulation in the area of the spinal vertebral lesion and in the physiological consequences [elsewhere]. . . . These uncertainties constitute a great obstacle to many scientifically trained men in maintaining an interest in osteopathy. Uncertainty regarding an observation to them means that it probably is not valid."[13]

While clinical research in distinctive osteopathic procedures was standing still, the value of new chemotherapeutic discoveries was steadily being demonstrated. In 1935, the first of the synthetically produced sulfonamides useful against hemolytic streptococci and staphylococci was introduced. Early in the 1940s penicillin, effective against the range of gram positive bacteria, became available. Beginning in 1945 streptomycin, which destroyed gram negative bacteria, was marketed. This was followed by aureomycin, chloramphenicol, and tetracycline. In addition to these antibiotics, between 1945 and 1960 a number of new analgesics, anti-inflammatory agents, muscle relaxants, and tranquilizers, as well as other forms of drug therapy were introduced.[14] The pharmaceutical manufacturers could provide tangible (if not always reliable) statistical evidence supporting the value and safety of their products, but the advocates of OMT could offer little more than testimonials. As a result, the "scientifically trained" DO to whom Chandler referred was likely to put greater trust in these modalities than in manipulation.

The relative decline in dependence on osteopathic manipulative treatment over the years is imperfectly reflected in the changing focus of the *Journal of the American Osteopathic Association.* In the early 1930s OMT was still included in the majority of articles and was described with great care and detail, but by 1948 the AOA Board of Trustees felt compelled to pass a resolution urging that "every effort be made by the writers of scientific papers for publication in the official Journal of the Association or in other osteopathic periodicals to include wherever feasible discussion of the relationship of the osteopathic concept to the subject of the paper."[15] However, as the contributors were increasingly specialists who had eschewed structural diagnosis and manipulative treatment in their own practice, this resolution had little if any impact. By the end of the 1950s most *JAOA* articles failed to mention OMT, and when they did it was only briefly and more as an adjunct than as an integral part of patient management. Articles devoted solely to osteopathic principles would still appear, but with far less frequency.[16] Those practitioners who did employ palpatory diagnosis and manipulative treatment increasingly referred to themselves as "ten-fingered" DOs as opposed to "three-fingered" DOs—the latter needing just that many digits to write a prescription.

Precisely how many DOs in the 1950s were performing manipulation and to what extent is impossible to determine, although one can come to certain general conclusions. Hospital-based DOs were utilizing OMT infrequently, and only a minority of surgeons saw to it that their patients received pre- and postoperative treatments. In office-based general practice,

OMT appears still to have been used with some regularity, but less time was devoted to it and it was increasingly restricted to the treatment of local joint and muscle problems. While approximately 10 percent of all active DOs, either through choice or because of state laws, exclusively employed distinctive osteopathic procedures, this group (comprised mostly of older DOs) was steadily shrinking each year. For younger practitioners, the data are more substantial. In 1954 the AOA mailed out a confidential questionnaire to all active DOs who had graduated between 1948 and 1953. Close to 60 percent responded. Only 44 percent of those answering the question "What percentage of your patients receive manipulative treatment?" responded, "over 50 percent." Considerable variation by school was noted; 53 percent of Kirksville graduates responded with this figure, compared to 16 percent of Los Angeles graduates.[17] Clearly, the DOs as a group were coming continually closer to the MDs in means of patient management.

Those DOs who strongly believed in the appropriateness of OMT for a wide range of conditions did what they could to alter this trend. In 1938 a group calling itself the Osteopathic Manipulative Therapeutic and Clinical Research Association, which after 1944 was renamed the Academy of Applied Osteopathy (AAO), was granted affiliate status with the AOA. The academy arranged programs focused on osteopathic principles and practice at the national AOA conventions, circulated papers to members, provided speakers for state and local AOA meetings, conducted short postgraduate refresher courses, and sponsored the writing or distribution of books relating to distinctive osteopathic approaches. Although these activities kept traditional osteopathy before the profession, the influence of the academy was limited. Only 12 percent of all DOs in the AOA were affiliated with the academy group at its peak.[18]

Other DOs viewed the AAO with skepticism, because it embraced members who made broad claims that ran counter to scientific facts or conventional understandings, the most controversial of which were promulgated by a Minnesota practitioner named William Garner Sutherland, DO (1873–1954). Based on his experiences, which dated from his training at the Kirksville School, Sutherland wrote and self-published a volume entitled *The Cranial Bowl* in 1939.[19] Sutherland argued that there was a "primary respiratory mechanism" which could be felt by placing both hands on the skull until the practitioner can gain the palpatory sensation of widening and narrowing. He believed that this primary respiratory mechanism consisted of five elements: the inherent motility of the brain and spinal cord, the fluctuation of the cerebral spinal fluid, the motility of intracranial and intraspinal membranes, the articular mobility of cranial bones, and the involuntary mobility of the sacrum between the ilia. Sutherland claimed that through palpatory techniques one could ascertain abnormal positions of cranial and adjacent structures which could be "guided carefully, gently, firmly, and scientifically into normal relationship."[20] Left alone, these "cranial lesions" could interfere with normal physiological processes throughout the body. Though this book had a largely mechanistic orientation, other of his writings and presentations were vitalistic in orientation. Sutherland believed that cerebrospinal fluid, membranes, and bones were propelled by an external force or energetic potency that he called the "breath of life." Sutherland's belief in the mobility of cranial sutures past infancy, as well as a number of his other ideas, ran directly counter to prevailing scientific evidence and opinion. He and many of his followers were noted for the zeal and enthusiasm with which they promoted his doctrines, which in turn alienated those DOs who were seeking external legitimacy for the profession. Eventually, Sutherland's followers formed the Cranial Academy, which became an affiliate of the AAO.[21]

STATUS INCONSISTENCY

The most vexing problem for the DOs as they expanded their scope of practice and improved their educational standards was the public's failure to recognize the complete range of services they could provide their clients; along with this oversight came a concomitant lesser deference and lower social standing than was accorded MDs. To the many DOs who believed themselves as well trained or as competent as their allopathic counterparts, these circumstances led to considerable frustration and alienation.

Part of their difficulty lay in their small numbers. From the turn of the century up through 1960, DOs constituted approximately 3–4 percent of the total U.S. physician population (MD and DO totals combined). Furthermore, as previously noted, the DOs were distributed disproportionately, and in many sections of the country osteopathic care was unobtainable. Indeed, as late as 1960, twenty-two states had fewer than fifty DOs apiece. This scarcity helped to make the profession socially invisible.[22]

Another handicap preventing widespread recognition and approval was their difficulty in securing the same legal privileges enjoyed by MDs. This applied not only to unlimited licensure but to winning the right to handle workers' compensation cases, becoming state or local health officials, entering the military medical corps, gaining access to public hospitals, and having their services covered under private insurance plans sanctioned by special enabling acts. Not having any or all of these rights and privileges served "officially" to brand the DOs as inferior practitioners.

Because of these various circumstances the great majority of Americans were unclear as to who the DO was and what precisely he or she did. In 1936 the AOA hired a public relations counselor, who conducted several on-the-street interviews in downtown Chicago. To the question "What is an osteopath?" a magazine writer responded, "An osteopath is a fellow who sets your spine, an MD who specializes in that method." A women's clothing stylist answered, "He's a man who has something to do with the spine." A bus driver declared, "Well, I don't know if I can word it. He massages people." Displaying an anatomical focus not surprising given his occupation, a postal clerk replied, "An osteopath has something to do with care of the feet." Another postal clerk asserted, "An osteopath has something to do with massage like a chiropractor. Osteopaths I believe are outlawed in New York and some other states too. I read about some of them being arrested in New York." A department store clerk exclaimed, "Oh yes, I know, I went to one once. They are especially for nervous people and treat them by massaging. The difference between a doctor and an osteopath is that an osteopath is drugless." A policeman reflected, "Let's see. He's a guy that when somebody gets all bent up, they put him on a table and twist him around and sorta put him together again. Ain't that right?" And finally, a stockbroker remarked, "He's a man who lays you on a table and massages. A doctor can be an osteopath but an osteopath can't be a doctor."[23] Although these beliefs were expressed in what was then a limited-practice state, the situation in states where osteopathic practice was unlimited or nearly unlimited was not much better. During the same year, Professor George Hartman of Columbia University published a study of the relative social status of twenty-five medical careers as judged by 250 Pennsylvania laypersons. The category "osteopath" was ranked eighteenth overall, one notch below "dietician."[24]

The profession acted in a number of ways to change its image. The first effort, which gained some momentum in the 1920s, concerned the matter of occupational title. Because the term osteopath had been so closely identified with manipulative treatment, many DOs believed new labels were needed. Dr. Alice Foley, writing in the *JAOA*, related the story of an attorney who said to her, "'Now you osteopaths do so and so, but the physician does thus and so.' The thought came to me that the public does differentiate. They call the allopath their physician and think of us as osteopaths." Foley recommended the use of the term *osteopathic physician*. "That explains the kind of physician we are, and it also leaves the word 'physician' in their thoughts concerning us." M. F. Hulett, DO, agreed, stating, "many of our friends are not yet aware of the fact that we are physicians at all, and some still seem surprised that we really treat the sick." Commenting upon *osteopathic physician* and *osteopathic physician and surgeon*, Hulett added, "I am quite sure the repeated use of these terms will add to the dignity of our school." This move received support from then-AOA editor Cyrus Gaddis, DO, who sermonized: "Let no piece of literature be circulated or none go out with simply 'osteopath.' Let it stand out 'osteopathic physician' and then be sure that we are ready to live up to that name. . . . Are you an osteopathic physician or just an osteopath? Times are changing. Are we willing to have the public consider us simple treatment givers?"[25] By 1940 the great majority of DOs were using the new labels.

Concurrent with the move by DOs to modify their shingles and office stationery, the national and state associations sought to bring up to date those references to *osteopathy* found in dictionaries and encyclopedias. Phrases or definitions which suggested that DOs were not in favor of drugs, or that they placed chief attention in their work upon finding and removing structural lesions through manipulation, were excised in favor of language that emphasized that osteopathy was a complete school of healing.[26] Telephone directory listings were also altered, substituting *osteopathic physician* or *osteopathic physician and surgeon*, depending on the licensure law, for the now increasingly discarded word *osteopath*.[27]

The AOA also worked to improve the DOs' social standing through the legislatures and the courts. As a result, many licensure laws would be revised in their favor; DOs became included in some of the health-related New Deal programs; they would win attorney general and judicial decisions on their participation in workers' compensation cases; they triumphed in some key battles over their right to access to hospitals built with public money; and they qualified for some federal aid, mostly in the form of Hill-Burton hospital construction grants.[28]

All of these efforts helped to improve the DOs' public image through the 1950s, but the rate of progress as perceived by many osteopathic practitioners was far from satisfactory. One of the principal reasons for their failure to make larger inroads in lay understanding was a consequence of their inability to convince the print media to notice them. National and state conventions, as well as DO speaker tours, were not considered particularly newsworthy; and when such events were reported, the resulting story was usually no more than a few paragraphs in length and was placed in an inconspicuous section of the newspaper. Most national general feature magazines also seemed to see little of interest in the profession, and those that did focused entirely upon the manipulative aspect. Typical was Mark Sullivan's "If I Need Relaxation," published in the *Reader's Digest*, an article that, while most complimentary, cast the DOs as highly skilled "rubbers" rather than broadly trained physicians, thus reinforcing the image the movement wanted to shed.[29]

When newspapers and magazines gave prominent attention to the activities of individual MDs, it was often in connection with the introduction of a new drug or a new life-saving surgical technique, or the reception of a prestigious award. Other MDs were regularly featured in periodicals by contributing health columns. The DOs, the great majority of whom were in the unglamorous field of general practice and made no spectacular contribution to research, were thus cut off from such favorable coverage. Indeed, when reporters focused on the exploits of individual osteopathic practitioners, it was almost invariably in connection with alleged or actual deviant behavior: a botched operation, an injury related to manipulation, an illegal abortion, a quack cure, or the like. When MDs were charged with similar malpractice, the public, given its knowledge of the medical profession as a whole, could dismiss these as isolated instances. However, because the same public had little or no prior knowledge of osteopathy, the activities of a few could easily be generalized to characterize the abilities or behavior of all DOs.

During the 1950s, the national press focused on one osteopathic practitioner, Dr. Sam Sheppard of Ohio, who in a highly sensationalized trial, was convicted of the murder of his wife, a verdict later reversed. Sheppard's saga became the inspiration for the long-running television show and, much later, the movie entitled *The Fugitive*, in which the doctor protagonist was portrayed as an MD. Though Sheppard was repeatedly referred to as an "osteopath" in articles and on television, most accounts also noted that he specialized in neurosurgery. This latter piece of information undoubtedly surprised many readers and listeners, who thereby learned that a DO could concentrate his or her efforts in a field other than manipulation.[30]

The continual lack of public awareness of who DOs were and what they did and their positive contributions to American health care generated a considerable degree of frustration among many members of the profession. Although most practitioners simply accepted the fact that a certain portion of their work consisted of answering questions relating to how much educational training they received, what their scope of practice was, and how precisely they differed from the MD, other DOs found this situation intolerable. This problem of poor public perception affected the DO's family as well. During social interaction, wives and children were always at risk of being put into the uncomfortable or embarrassing position of having to explain or even defend their spouse's or parent's occupation. In 1955 AOA editor Dr. Raymond Keesecker, addressing the student doctor's wife, noted that such situations "Give you the best opportunity in the world for some important public relations work." However, some wives saw this as a terrible burden, and even Keesecker admitted that in dealing with any question about osteopathy, "it is not too easy to give a specific answer."[31]

Many DOs came to believe that the primary cause of their identity problem was the letters behind their names. The American public, they argued, recognized the MD degree as the universal symbol for a physician and surgeon; thus it was not all that surprising that patients seeing any other designation would be confused as to its meaning, even if the title *physician and surgeon* were added to their stationary or their shingle. In the opinion of some DOs, the easiest way of changing their image was to change the degree awarded by osteopathic colleges to that of MD. During the 1920s and 1930s, such calls were occasionally sounded in the journals, but they gained no support within organized osteopathy as a whole.[32] However, with America on the verge of entering the Second World War, some students and alumni of two of the schools pleaded with their administrations to adopt the MD designation, believing that through this maneuver they could become eligible to serve their country as military physicians. To take the onus off the school officials and put an end to such hopes, the AOA Board of Trustees in 1941 declared that "the only degree to be issued by an approved osteopathic college qualifying for licensure to practice the healing art shall be the degree of doctor of osteopathy."[33] As far as the AOA was concerned, this decision was absolute and irrevocable.

The refusal by the AOA to accommodate this dissatisfied minority led some DOs to obtain diploma-mill MD degrees to hang in their offices. Such certificates, while totally worthless for the purposes of licensure, were nonetheless thought useful by their possessors as a means of convincing new patients that they were after all "real doctors."[34] However, a larger group of unhappy practitioners were not willing to go that far. They simply decided to leave all mention of their DO degree and reference to osteopathy off their stationery and shingle and just go by the title "Dr. 'So-and-So', Physician and Surgeon."[35] Thus, while the general public was confused as to what DOs did, a significant number of osteopathic practitioners were coming to the conclusion that the best way to deal with the confusion of laypersons was to hide their identity.

8

THE CALIFORNIA MERGER

For most DOs the problems connected with their identity did not undermine the desire for professional autonomy. Even many of those who failed to advertise themselves as osteopathic practitioners and favored the schools' awarding an MD degree continued to believe they were part of a distinctive group that should remain politically separate and independent. Their displeasure with the AOA was with its policy, not its legitimacy as the voice of osteopathy. However, this attitude was not universally shared. For some DOs the various changes taking place within the profession, combined with their specific situation at the local level, led to a vastly different interpretation and outlook. Nowhere was this more evident or widespread than in California.

THE COA AND THE CMA

In the decades prior to 1960 there were more DOs practicing in California than in any other state; they constituted at any given time 10 percent of all its physicians, and perhaps 15 percent of its total population were their patients.[1] In terms of legislative victories, public acceptance, and average practice income, no other state group approached their achievements; however, a deep disenchantment with their lot belied their outward success.

Even before the turn of the century California DOs were beginning to establish themselves as the progressive wing of osteopathy. In 1896 the Pacific College became the first school to introduce a mandatory two-year course and later was one of the earliest to expand to three years. The state attracted and became the stronghold for the "broad osteopaths," and under their influence the Los Angeles College became the first institution to place in its curriculum a course on *materia medica*. The College of Osteopathic Physicians and Surgeons (COP&S)—the result of combining the aforementioned schools—was the first to require one year, then two, and ultimately three years of prior college work as an entrance requirement. Finally, with respect to clinical facilities, it was the first and only school to utilize a large municipal hospital for bedside and outpatient teaching.

The Los Angeles County General Hospital-Unit #2 trained many of the profession's leading specialists and was well regarded in the community for the quality of care that patients received. As it was a government institution, the hospital administration published annual statistics, and those numbers appeared to demonstrate year after year that the mortality rates and length of patient stay (every tenth patient was admitted to the osteopathic unit) were consistently better than the much larger Unit #1 next door, run by the MDs. The MDs claimed that a significant number of sicker patients were transferred or diverted to their unit, a charge the DOs dismissed. As a result of these continuing unfavorable comparisons, the MDs pressured the hospital administration in 1934 to change the rules regarding admissions and publish only statistics that represented the combined totals of the two units.[2]

Pointing with pride at their various achievements, California DOs generally regarded themselves as the best qualified osteopathic physicians and surgeons in the country. Most also considered themselves the most "scientific," which meant that the decline in distinctive osteopathy was more pronounced in California than elsewhere. Furthermore, because of the stringent requirements of California law as well as tougher regulations adopted by the state board of osteopathic examiners, graduates from other DO schools were for many years ineligible for unrestricted licenses. This made for a comparatively homogeneous osteopathic population.

California DOs placed a high priority on securing for themselves complete equality with the MDs in their state. It was thus a source of continuing frustration for them that whatever progress they collectively made, significant gaps between the two groups remained. This was reflected most clearly in the matter of college finances. From the 1930s through the early 1960s, COP&S could raise and spend only one-half to three-fifths the amount MD schools could allocate for the education of each of their students. Unlike MD-granting schools, it could not count upon state support, general university funds, and philanthropy. Although COP&S employed more full-time faculty than other osteopathic colleges, it could not approach the numbers characteristic of an AMA-accredited institution.[3]

Far worse was their postgraduate situation, particularly with respect to training facilities for residents. Only a small number of positions were available annually within the entire state, and the great majority of those were offered at Los Angeles County General Unit #2. While several dozen DO facilities were founded in California after 1930, the bed capacity of most was too limited to provide for adequate specialty programs. Those applicants not successfully placed had to either go out of state or resign themselves to a general practice. At the same time, California DOs could only look with envy at the number and quality of allopathic hospital residencies in their midst, appointments for which they as osteopathic physicians were ineligible.

Added to these educational problems was lack of visibility. Public knowledge of the DOs, particularly the scope of their services, was far less than what was known of the MDs, in spite of the fact that a significant minority of the California population was served by osteopathic physicians and surgeons. Part of this problem may be attributed to the continuous waves of new residents arriving from areas of the country where the profession had far fewer representatives and limited practice rights. Some California DOs argued that in one sense they were being victimized by the national image of osteopathy, and they held the AOA responsible, alleging that it was too tolerant of the lower standards maintained by the other colleges and was not working hard enough to eliminate remaining legal and social inequalities elsewhere.[4] This combination of elements— namely one group of DOs thinking of themselves as a breed apart from the rest, added to the other problems that faced the profession generally, such as poorer educational opportunities, lack of public recognition, and a decline in the use of distinctive osteopathic procedures—led an increasing number of California practitioners to consider the possibility and advantages of leaving organized osteopathy for organized medicine.

While more California DOs were coming to this conclusion, so too were the leaders of the state's medical association, but for entirely different reasons. For the most part, organized medicine within California saw the DOs as an inferior group of practitioners who were lowering the general quality of health care in the state. For decades they had tried through various legislative means to eliminate the profession, but to no avail. Despite their small number, the DOs had been able to wield considerable political power, effectively blocking the passage of threatening measures. When in 1942 the California Medical Association supported a ballot initiative to create a basic science board, the DOs along with the chiropractors led a vigorous campaign against it. The proposal lost by a two to one margin.[5] With no other viable strategy left open to them, California MDs came to believe that the only way to destroy osteopathy was through the absorption of the DOs, much as the homeopaths and eclectics had been swallowed up early in the century.[6]

In 1943 Forest Grunigen, DO (1905–1999), president of the California Osteopathic Association (COA), appointed what was called the Fact-Finding Committee, to meet with representatives of the California Medical Association (CMA) following five years of informal contacts concerning merger.[7] At this first official meeting the CMA offered a proposal for amalgamation which they had already discussed with the AMA Council on Medical Education and the Association of American Medical Colleges. This plan called for the granting of MD degrees, by one of the existing four medical schools in the state, to all DOs licensed as physicians and surgeons in California; the elimination of the osteopathic licensing board; and the conversion of COP&S into a medical school.[8]

In February 1944, the COA committee was advised by its counterpart that both the council and the American Association of Medical Colleges had given their tentative approval to the outline of their plan. However, within a month the Federation of State Medical Boards announced that it would refuse to recognize the validity of this MD degree and warned that any medical college issuing such a diploma would lose the right to have any of its regular graduates examined for licensure. Strong opposition was also voiced by certain influential AMA leaders, notably Morris Fishbein, MD, editor of the *Journal of the American Medical Association*, who was ready to fight any accommodation between "physicians" and "cultists."[9] These moves forced the American Association of Medical Colleges and the AMA Council on Medical Education to back away from their proposal. At the spring 1944 COA convention, the head of the Fact-Finding Committee noted that any possibility for amalgamation in the near future had disappeared.[10]

Staff members at the AOA headquarters in Chicago as well as key national leaders were kept apprised through their local contacts of the events in California. Their strategy seems to have been to do and say nothing. First, they did not want to be viewed as interfering in the affairs of the organization's largest component society; if they were to do so, those who favored a merger might capitalize upon the issue and gain support. Second, they believed that the merger talk would probably lead nowhere. And third, they believed this effort to be triggered by the desire to obtain an

MD degree just to be eligible to serve in the military medical corps. Once victory overseas had been achieved, they figured, sentiment in this direction would undoubtedly lessen. As a result of the AOA decision not to print any information or commentary in its journals, most DOs outside California remained ignorant of the entire matter.

In the late 1940s, however, the attention of the profession was focused upon California when a group of dissident DOs set up their own "medical college" for the purpose of granting "academic" MD degrees to any osteopathic physician and surgeon who paid his or her tuition fee and attended thirty-six hours of lectures. During 1947 and 1948, at least 137 DOs secured one of these MD degrees. Graduates of this institution, known as Metropolitan University, then set up what they called the Pacific Medical Association, which began lobbying for its own legislative program.[11] Both the COA and the AOA took a strong stand against these activities. In 1948 the AOA House of Delegates unanimously amended the code of ethics to prevent any DO from possessing or displaying any unaccredited degree, and the following year, through national and COA pressure, both the Metropolitan University and the Pacific Medical Association were forced to disband.[12]

The position taken by the COA in this matter seemed to convince many of those AOA leaders acquainted with the merger attempt of a few years earlier that such a threat had passed. However, the action of the COA was primarily motivated by its belief that the Metropolitan people, with their worthless degrees, were embarrassing the profession and that the establishment of a rival lobbying group would only sap its own political strength. In fact, the desire for a merger by some COA leaders and many California DOs in the field had not diminished. In 1947, Dr. Grunigen was appointed head of the COA Fact-Finding Committee, and he and his committee members over the next fourteen years held formal and informal discussions with CMA representatives on how this goal might be achieved.[13]

THE AOA-AMA CONFERENCE COMMITTEE

Both DO and MD discussants in these continuing COA-CMA talks came to recognize that as long as the medical profession generally and AMA officials in particular held a decidedly negative view of osteopathy, a merger would be most difficult, if not impossible, to arrange. One possible way to break down this hostility would be to get their respective national leaders to hold conferences about common concerns. Such interactions could very well lead to a better understanding between the two larger associations and an upgrading of the status of the osteopathic profession, which in turn would facilitate their local efforts. Since both MDs and DOs from California held significant leadership positions in their respective national associations this plan was soon followed.[14]

In 1949 the COA House of Delegates, through its representatives, urged the AOA Board of Trustees to establish a fact-finding committee that would be prepared to meet with any healing arts group. The Californians, notably Grunigen, who was then first vice-president of the AOA, pointed out that in recent years their own Fact-Finding Committee had been able to resolve differences with the CMA and even with other state health associations over proposed legislation and had reduced mutual mistrust. If any of the AOA board members were skeptical as to the motivation behind the California proposal, they kept their doubts off the public record. Instead, the board quickly and quietly approved the measure, although the mechanics of how such talks might be initiated, particularly with the AMA, remained unresolved for more than a year.[15]

In October 1950, Floyd Peckham, DO, of New York, president-elect of the AOA, addressed the Kentucky Osteopathic Association. At the instigation of a DO member of the state board of health, he was able to meet informally with AMA president Elmer Henderson, MD, a local resident. During their friendly chat, a legal problem faced by the Chicago College of Osteopathy was raised. Although its graduates were then eligible for unlimited licensure in thirty-five states, this was not true in Illinois itself. As a result of their talk, both agreed that a conference should be arranged between representatives of their associations to discuss Illinois licensure in greater depth.[16] In December the AOA board appointed a committee of five DOs, including two Californians, which met with a similar group from the AMA in February 1951. One of the latter committee's members was John W. Cline, MD, of California, president-elect of the AMA, who had played an advisory role in furthering the CMA amalgamation proposal eight years earlier.[17]

At this meeting the MDs were of one mind that the question of the Chicago school was a matter for local, not national, action; and so, with the DOs acquiescence, the matter was dropped. Some of the MDs brought up the issue of amalgamation of the two professions, but this was coldly received. In all, nothing of substance was accomplished. Yet, despite this lack of agreement on topics to be discussed, it was the feeling of the participants that future conferences might prove useful.[18]

That June, Cline, newly installed as AMA president, reported to his Board of Trustees that "the relations between medicine and osteopathy present . . . widespread problems involving a majority of the states to some degree," and he therefore urged that it appoint a committee for the second time to discuss these matters with AOA representatives.[19] The board acceded to this request, and soon afterwards a similar committee was appointed by the AOA. At this conference, held in March 1952, discussion centered on the question of the longstanding AMA position designating the DOs as "cultists." Obviously, this was a stumbling block that had to be overcome if the two associations were to resolve other problems, particularly the matter of interprofessional consultation and the issue of DOs and MDs serving on the staffs of public hospitals. This meeting ended with a greater degree of understanding on both sides.[20]

In his farewell presidential speech in June 1952, Cline, addressing the AMA House of Delegates, briefly mentioned the work of the Conference Committee, declaring that osteopathy had in recent years come much closer to medicine and that "removal of the stigma of cultism would hasten that process." As a first step, Cline recommended that the Council on Medical Education and Hospitals be permitted to aid and advise osteopathic schools, and that any ethical barrier now preventing MDs from teaching in these colleges be removed. No action was taken on Cline's suggestions, though the board agreed to let its Conference Committee meet again with the AOA "when or if requested." Meanwhile, the AMA Judicial Council was asked to prepare an opinion.[21]

Up until this time, the Conference Committee discussions had taken place without the knowledge of the rank and file of either profession. However, with Cline's address the meetings had become public knowledge. In some osteopathic circles this news was interpreted as meaning that the two associations were conspiring to arrange a merger, which in turn caused an outpouring of angry letters, telegrams, and phone calls to the AOA headquarters from outraged DOs across the country. In successfully allaying their fears, the AOA board and house in July 1952, reaffirmed "in the strongest terms possible the policy of maintaining a separate, complete and distinctive school of medicine." This was followed up by editorials in AOA publications explaining the history of the conference committees and the content of the discussions to date.[22]

In December 1952, the AMA House of Delegates was informed by the Judicial Council that it had no report from the Conference Committee that had been appointed in June. Lacking any additional information on osteopathy, it could do nothing more than "reassert its opinion that all voluntary associations with osteopaths are unethical." Cline, who had been made head of the Conference Committee, pointed out in response that the wording of the June resolution, namely that they would meet "when or if requested," was holding up future talks. As a result, the AMA removed the troublesome precondition, giving Cline a free hand. The AOA board in response gave its committee a similar charge, and another meeting was scheduled.[23]

In preparation for this third session, Cline set about collecting what historical and current data he could find on osteopathy—its schools, hospitals, and laws. He and other committee members consulted with various DOs and the AOA central office staff and sent out questionnaires to osteopathic colleges as to the elements of osteopathic education. When the two conference committees met in May 1953, they jointly reviewed the information the MDs had gathered, and the AOA representatives furnished further materials.[24]

The following month Cline presented a detailed report on osteopathy to the AMA Board of Trustees for their consideration. In it he declared that while the original teachings of Andrew Taylor Still "could be classified as 'cultist' healing," a great evolutionary change had since taken place within the profession. While he noted that he was unable historically to trace this progress due to an absence of adequate secondary sources, Cline observed that over the previous several decades osteopathic colleges had fully integrated pharmacology, surgery, and all other orthodox modalities into their curricula and had reduced the time allocated to distinctively osteopathic features. Furthermore, the "osteopathic concept" or philosophy had been broadened. Though DOs had differences of opinion among themselves in this regard, the concept consisted of three basic principles: first, the normal body contains within itself the mechanisms of defense and repair in injuries resulting from trauma, infections, and other toxic agents; second, the body is a unit, an abnormal structure or function in one part exerts abnormal influence in other parts; and third, the body can function best in defense and repair when it is in correct structural alignment. Cline went on to describe the colleges in terms of their facilities, class sizes, curricula, and quality of instruction. He noted the geographic distribution of DOs, the volume of care they delivered, the scope of licensure, postgraduate education, and the current state of relationships between MDs and DOs.

In concluding his presentation, Cline and his committee made four recommendations: (1) that the House of Delegates declare that, as so little of the original concept of osteopathy remained in the way medicine was currently taught in osteopathic schools, it could not be classified as the teaching of cultist healing; (2) that it be the policy of the association to encourage improvement in undergraduate and postgraduate education of doctors of osteopathy; (3) that the state medical associations determine for themselves whether professional relations between MDs and DOs were ethical; and (4) that the Conference Committee be established on an ongoing basis. After mulling the matter over, the AMA board decided, because of the length of the report and the controversial nature of the subject, that the House of Delegates would need further time for its study and, in addition, that the component state medical associations should have the opportunity to express their opinions. The committee was continued, but action on the report was deferred for one year until June 1954.[25]

The following September, Floyd Peckham, DO, head of the AOA Conference Committee, telephoned Cline to express appreciation for his efforts and to find out if there was anything more he could do to assist in removing the cultist label. Cline, in response, suggested the possibility of on-campus visits by the committee to osteopathic schools, explaining that the most telling criticism of his report was that his information was secondhand and hearsay. While he himself knew the data to be reliable and his statements factual, this did not satisfy the skeptics, and many of these individuals would have to be won over if his recommendations were to have a chance of passage next year. An on-campus visit by the committee, accompanied by distinguished medical educators, would help to undermine the opposition.[26]

In October the AOA Conference Committee formally met and Peckham conveyed the substance of his conversation with Cline. The committee agreed that they would give his plan due consideration when it was submitted, but that this was a matter to be decided by the Board of Trustees. Early in December, following the annual AMA house meetings, Cline telephoned Peckham with the news that he had cleared his proposal with all the necessary authorities within the AMA and could provide the details of his plan, which he did in a letter dated December 8, 1953. As a pattern for the visits, Cline suggested the same type of unfocused, comprehensive survey routinely carried out by the Council on Medical Education and Hospitals for the purposes of accreditation. Peckham immediately realized that this would be unacceptable; nevertheless, a special session of the AOA board was convened for a hearing. As expected, the Cline proposal was formally rejected; however, the board decided not to preclude the possibility of any on-site visitations per se. It instead directed its Conference Committee to meet with its counterpart to see if they could agree on a satisfactory compromise.[27]

On January 16, 1954, the two conference committees again met, whereupon the DOs listed a number of conditions they believed would facilitate approval of a visitation. The conditions included: (1) language within the AMA proposal stating clearly that their on-campus inspection would have nothing to do with accreditation and affirming that the AOA Bureau of Professional Education and Colleges was the only authoritative body that had the right to accredit osteopathic schools, and that under no circumstances was it the will of that committee to disturb or upset that responsibility; (2) wording to the effect that the primary purpose of the visitation was to determine whether or not "medicine as currently taught in schools of osteopathy constitutes the teaching of 'cultist healing'"; and (3) the establishment of the right of the AOA to reject any of the visitation advisors proposed by the AMA. Cline and his committee immediately accepted all of these conditions.[28]

The following month another special meeting of the AOA board was convened, which was attended by a representative from each osteopathic college as well as by other key members of the profession. Cline's revised proposal, despite integrating all of the suggestions made by Peckham's committee, still met with serious objections. Some board members were afraid that the resulting report would be negative and therefore might seriously harm the profession in its legislative efforts. They had not forgotten the problems caused by the Etherington-Ryerson survey of two decades earlier. There was also fear of the outcome should the report be favorable. This conclusion might signify to lawmakers that the need for independent osteopathic licensure boards had passed and in this way give impetus to the push for their elimination. Furthermore, a positive report might lead the AMA not only to remove the cultist label but to launch a campaign to bring about complete amalgamation. These doubts led the board to defer immediate action and to put the matter before the next regular session of the House of Delegates, which was to meet one month after Cline's original recommendations were to be voted upon.[29]

At the AMA board and house meetings in June, Cline won approval of another year's delay, based on the prospect of an AMA inspection of osteopathic schools in fact taking place. This left the decision to the DOs. During its session in July the AOA house debated the same issues that had come up at the February board meeting. This time, however, the members of the Conference Committee took a far more active role in the discussion, arguing that unless the AMA visitation was approved, there was no chance that the cultist label would be removed. The argument proved persuasive. The opposition was overcome and the Conference Committee was given full authority to negotiate with its counterpart in making final arrangements.[30]

THE AMA INSPECTION

Under the terms of the agreement between the two associations, each osteopathic school had the right to decide whether or not it would participate in the inspections. By late October five of the six colleges had given their approval; the Philadelphia school argued that the visitations still looked too much like an accreditation process and therefore declined. As the original AMA mandate called for a survey of all six colleges, Cline's committee had to wait until the AMA clinical meeting in December to receive official authorization to inspect only five.[31]

Prior to the visitations, which took place between January and March 1955, each participating school filled out a questionnaire patterned after that required of colleges seeking accreditation by the AMA Council on Medical Education and Hospitals. The inspection team requested essential information concerning organization, authority, administration, finances, facilities, and operation of the colleges; the personnel, training, authority, and activities of the faculty; the curriculum content; the organization of departments, their objectives, methods of teaching, and equipment; the degree of interdepartmental coordination and cooperation; and the details of library facilities and contents.

Each institution was visited by at least two members of the committee, which then consisted of Cline, James Appel, MD, Leonard Larson, MD, Thomas Murdock, MD, and Cleon Nafe, MD, accompanied by one of the mutually agreed upon educational advisors: L. R. Chandler, MD, recently retired dean of the Stanford University School of Medicine, J. Murray Kinsman, MD, dean of the University of Louisville School of Medicine, and W. Clarke Wescoe, MD, dean of the University of Kansas School of

Medicine. Floyd Peckham accompanied each team. It was agreed beforehand that the inspection committee would have access to all the information they believed essential to their efforts and that the observations would be of such breadth, depth, and duration as they deemed necessary. At the end of each on-site visit, the advisor prepared a report, one copy of which was transmitted to the college, while the other was held by the AMA committee as a confidential document. Following completion of all visitations, the committee drafted a final document containing the answers to four questions posed to it by the AMA board: (1) Is modern osteopathic education the teaching of "cultist" medicine within the definition of the principles of medical ethics? (2) If the first question is at all true, to what degree? (3) If to some degree, does this element interfere with sound medical education? (4) What is the quality of medical education?

In its findings, presented to the AMA House of Delegates in June 1955, the committee noted that all of the schools were attempting to give their students a rounded general practitioner–type of training, expecting that the majority of their graduates would become primary care physicians and that a high percentage would locate in traditionally underserved communities. Examining student records, the committee observed that all students had completed the educational requirements for admission to an AMA-accredited college and that a considerable number of them could have obtained admission to medical school. Interviews conducted with students revealed that the motivation to become a physician was strong in most. While some were disappointed medical school applicants, more had previous contacts with the osteopathic profession and were thereby influenced to enter DO institutions. A small number, in fact, had been accepted by MD colleges but had chosen an osteopathic school instead.

All of the schools, the committee observed, were handicapped by limited finances; endowments were small or nonexistent, and too much of their funding was derived from tuition. Because of this financial situation, the schools were not able to hire more full-time faculty and improve their facilities and equipment to the extent that the colleges would have liked. Though the committee noted that in recent years additional sources of funding (for example, the Osteopathic Progress Fund, federal teaching grants, and Hill-Burton monies) had allowed the schools to make some significant changes, considerably more support was necessary.

In terms of curriculum, the committee found that the clock hours of osteopathic instruction exceeded those of schools of medicine by several hundred. This, it felt, was not advantageous to the student, since it crowded too much into too limited a period. As a result, there was insufficient time for individual student projects, library use, and reflection and assimilation of the knowledge the student acquired. Furthermore, the situation did not encourage a scholarly attitude or an interest in research.

In the basic sciences, the committee concluded, subjects were fairly well taught and the students were well grounded in these fields. Some departments—most frequently anatomy—were outstanding, although some, particularly pathology, were comparatively poor due to a shortage of trained personnel. In the clinical years, the committee believed, there was too much didactic teaching and a tendency to treat the student as an observer rather than as a part of the patient care team. The methods and quality of clinical instruction, it found, varied from school to school, and to a considerable degree within different courses in the same college. A similar finding was made of the qualifications, teaching abilities, and interests of the faculty members. Finally, the committee felt that, in a majority of the colleges, the clinical material available was inadequate for the number of students.

On the most controversial aspects of osteopathic instruction, the committee believed, what was being taught simply reflected a difference in emphasis in both theory and practice between MDs and DOs, rather than a conflict between science and nonscience. What DOs referred to as the "osteopathic concept" was merely the expression of these differences. "Modern osteopathic education," the report noted, taught "the acceptance and recognition of all etiological factors and all pathological manifestations of disease as well as the utilization of all diagnostic and therapeutic procedures taught in schools of medicine."[32]

In the committee's view, the curricula had relegated osteopathic manipulation to the status of an adjunct to therapy within the sphere of medicine. Nowhere did it occupy a preeminent place in instruction. When applied to hospital inpatients with clinically recognized disease, for example, the committee found that manipulative treatment consisted mainly of relaxing, soft tissue manipulation or that designed to increase respiratory excursion. Some heads of clinical departments believed that OMT had considerable value when used in conjunction with standard therapy, while others did not. "The use of manipulative therapy," the report observed, "is decreasing in colleges of osteopathy and is increasing in the orthopedic and physiatry departments of medical schools."[33]

At the conclusion of its report the committee restated, though in somewhat revised form, the recommendations originally submitted in 1953. It urged the house to declare that current education in osteopathic colleges did not constitute the teaching of cultist healing; that MDs be encouraged to assist in osteopathic pre- and postgraduate training programs in those states where such participation was not contrary to the announced policy of the state medical association; that these same state associations assume the responsibility of determining the ethical relationship between MDs and DOs or request that their component county societies to do so; and that the Conference Committee be continued to meet with AOA representatives concerning common or interprofessional problems at the national level.[34]

Upon submission, the report was sent to the AMA Reference Committee on Medical Education and Hospitals, which in turn presented a majority and minority opinion. Both declared that, unlike the inspection team, they were not satisfied that current education in DO schools was free of the teaching of "cultist healing." However, beyond this the committee members sharply differed. The majority report, representing four out of the five members of the Reference Committee, urged the passage of the Conference Committee's last three recommendations. The minority report, consisting of the views of one member, Milford Rouse, MD, of Texas, urged rejection of all four recommendations and the adoption of two substitutes: first, that the Conference Committee be thanked for its diligent work and be discontinued, and second, "that if and when the House of Delegates of the American Osteopathic Association, its official policy-making body, may voluntarily abandon the commonly so called 'osteopathic concept,' with proper deletion of said 'osteopathic concept' from catalogs of their colleges and may approach the Board of Trustees of the American Medical Association with a request for further discussion of the relations of osteopathy and medicine, then the said Trustees shall appoint another special committee for such discussion." After a vigorous and emotional debate on the floor of the house, the motion to adopt the majority report was amended to substitute the minority report in its place. Upon further discussion, the house by a vote of 101 to 81 passed the Rouse resolutions. All the findings of the college inspection team were thus repudiated, and the DOs officially remained "cultist" in the eyes of the AMA.[35]

AN AMALGAMATION IN CALIFORNIA

The reaction of the AOA Conference Committee members who met two days after the AMA house vote was one of bitter disappointment. Although they were far from pleased with "the Cline Report" in its entirety, they concluded that on the whole it was reasonably fair. As far as the decision of the AMA house was concerned, it simply reinforced their belief that politics was at the heart of the cultism issue.[36] At the AOA House of Delegates meeting the following month, the actions of the AMA were largely ignored in the official sessions. The association restated its position of cooperation with any group whenever such cooperation might be expected to lead to improved health care for the American people. It retained its national Conference Committee for that purpose and urged the establishment of similar committees on the local level.[37]

As in the case of their national association, leaders of the California Osteopathic Association were greatly upset by the AMA house vote on the Cline Report. Nevertheless, they clearly recognized that the on-site visitations had opened many of the delegates' eyes to what was actually taught in osteopathic schools and in so doing had helped to raise DOs' standing and status among MDs generally. What now remained to be done to facilitate a local merger was to push the AOA towards meeting the Rouse conditions, so that the stigma of the cultist label would be removed, thereby eliminating any possible AMA objections to amalgamation. At its May 1957, meeting the COA House of Delegates passed a resolution urging the deletion from all AOA printed materials of those statements referring to the osteopathic profession as a separate, independent, and complete school of medicine, and the removal of all possibly "cultist" terminology employed by the colleges and hospitals, and it directed its delegation to the AOA house to make every effort to implement these changes.[38]

That July, during the national convention, debate centered on the AOA constitution, which then read in part:

> The objects of this Association shall be to promote the public health and the art and science of the osteopathic school of practice of the healing arts, by maintaining high standards of osteopathic education and by advancing the profession's knowledge of surgery, obstetrics, and the prevention, diagnosis and treatment of disease in general; by stimulating original research and investigation, and by collecting and disseminating the results of such work for the education and improvement of the profession and the ultimate benefit of humanity; that the evolution of the osteopathic principles shall be ever growing tribute to Andrew Taylor Still whose original researches made possible osteopathy as a science.[39]

In place of this awkward, "cultist"-sounding testament, a majority of the members of a Special Reference Committee of the house proposed that this part of the constitution, known as article 2, be amended to read simply, "The objects of this Association shall be to promote the public health, to encourage scientific research and to improve high standards of medical education." This was moved by the California delegation from the floor. A minority report, supported by representatives from Michigan, the second largest delegation, strongly argued that this statement led to questionable interpretations and argued substitution of the term *osteopathic education* for *medical education*. A seemingly certain bitter floor fight was narrowly averted when both sides agreed to compromise language: *medical education in osteopathic colleges*. This change was approved for publication and set for final action at the next year's meeting, where the house overwhelmingly approved it.[40]

At the December 1958 meeting of the AMA house, the Indiana delegation, following the Cline recommendations, again proposed that the state societies be given the responsibility for determining whether relations between MDs and DOs were ethical. This was rejected once more; however, the committee studying this resolution made the suggestion, which was approved, that the Judicial Council consider this matter further and submit a report.[41] During the next house meeting, in June 1959, the Judicial Council, specifically citing the recent AOA constitutional changes, now proposed a significant revision of association policy, recommending, "It shall not be considered contrary to the Principles of Medical Ethics for members of the AMA voluntarily to associate professionally with physicians other than doctors of medicine, who are licensed to practice the healing art without restriction and who base their practice on the same scientific principles as those adhered to by members of the AMA [and for AMA members] to teach students of osteopathic medicine who seek to develop and improve their ability to provide a better quality of medical care."[42]

To the surprise of many, this recommendation was opposed by the California delegation, which argued that the changes were too generous and which announced publicly that the CMA was then actively involved in negotiations with the COA to amalgamate the two professions and take over the osteopathic college. If the AMA were to give DOs all they asked for now, the bargaining position for the CMA would be weakened. On the AMA house floor the Californians led a successful fight to amend the entire resolution to read, "It shall not be considered contrary to the principles of medical ethics for doctors of medicine to teach students in an osteopathic college which is in the process of being converted into an approved medical school which is under the supervision of the A.M.A. Council on Medical Education and Hospitals."[43] Needless to say, the eyes of organized osteopathy now turned towards California.

The next month, at the AOA House of Delegates meeting, retiring association president George Northup, DO (1915–1996), focused on what appeared to be happening. In his address, Northup reviewed in some detail the discussions held in the early 1940s pertaining to merger and noted that, since then, rumors had periodically circulated that further talks along these lines were being held. Now, with the public statement by the CMA representatives that a merger between the two California groups was imminent, some clear and straightforward answers were due the AOA by its divisional society. Northup forcefully stated:

> In fairness to the remainder of the profession, its educational system, and its programs for the future, this profession and the House of Delegates has the right, yes, the responsibility to know whether there is any validity in these statements so that the AOA can act accordingly. If we are about to lose one of our prominent and best qualified colleges, we should face the possibility fairly and honestly. If the largest divisional organization of this profession is conducting through its leadership, official or unofficial, private negotiations with one of the largest divisional medical societies which might lead to the loss of their membership in the AOA that too must be faced realistically and honestly.

Northup then asked the entire House of Delegates four questions:

> (1) Do we wish to maintain the independence of our colleges or do we desire to convert them into medical schools under the supervision and jurisdiction of the Council on Medical Education and Hospitals of the AMA? (2) Do we wish to take steps leading to the abandonment of our intern and residency training programs, our approved and registered hospitals; our certification and recognition of our specialists and their certifying programs; our program of development and recognition of our general practitioners; and our hard earned acceptance of the AOA as a recognized accrediting agency, or are all of these to be turned over and placed under the protective custody of agencies of the AMA? (3) Do we or do we not have a contribution to make to medicine not now being accomplished through the efforts of any other organization? (4) Do we wish to continue as an independent osteopathic profession, cooperative with all and subservient to none?[44]

Northup's questions were answered with demonstrations of loyalty from seemingly all present except the California delegation, who sat, angry and silent, refusing to offer any explanation for their position. Michigan's delegation then introduced a resolution in direct opposition to the California House of Delegates' policy statement of 1957, reading in part, "Be it resolved that the osteopathic school of medicine, in the interest of providing the best possible health care to the public, shall maintain its status as a separate and complete school of medicine cooperating with all other agencies and groups that sincerely promote the same objective when that cooperation is on an equal basis granting full recognition to the autonomy and contribution of the osteopathic school of practice."[45] This passed 95–22, with California delegates dissenting.

Despite the AOA resolution, secret negotiations between the leaders of the COA and the CMA continued apace.[46] When in early 1960 word of these talks filtered back to AOA officials, a full accounting was demanded. At the July 1960 AOA house meeting, the Californians asked for and received permission to present their case before a closed-door executive session. Drs. Dorothy Marsh and Nicholas Oddo reviewed past differences with the AOA over legislation and setting of standards, noted the problems of obtaining adequate postgraduate training, observed the inadequate financing of all phases of osteopathic education, the poor status of the DO degree, and the exclusion of DOs from group health insurance plans. The profession's organizations, they maintained, were simply not moving fast enough to resolve these problems. Through amalgamation, these difficulties could be eliminated.[47]

DOs who opposed the merger argued that these various problems could be successfully dealt with in other ways. The Californians, they declared, simply wanted a quick fix, and in the process they were even willing to sell out their heritage. On returning to open session, the house resolved: "That any divisional society which is in the process of negotiation leading to unification and/or 'amalgamation' or merger, or a process of a similar nature, of the osteopathic profession with or into any other organized profession involved in health care shall cease such negotiations or be subject to the revocation of its charter by the AOA."[48]

Now for the first time under a direct threat, the COA leadership notified its members that it was instructing its Fact-Finding Committee to cease its discussions with the CMA. However, on November 13 the full COA house, in a defiant mood, voted 66–40 to ignore the AOA directive and resume talks. The AOA board reacted quickly. Meeting in special session the following week, it voted 18–1 to revoke the COA charter.[49] This left the COA in a precarious position, particularly if a merger agreement could not be worked out. In early December a new group, known as the Osteopathic Physicians and Surgeons of California (OPSC), was organized by loyalists and was quickly chartered as the official AOA divisional society.[50] Hopes by the OPSC leadership that it would soon represent over half the DOs in the state soon proved unrealistic. Only about one-sixth of all California DOs joined its ranks. The AOA decision to remove the COA's charter had the unanticipated effect of increasing the social solidarity of most California DOs and strengthening their identification with their established state society.[51]

By May 1961 a contract between the CMA and COA was ready to be acted upon by each house of delegates. The executive vice-president of the AMA had already assured the CMA that if unification was effected, "it would not be reviewed by any board or agency of the AMA for the purpose of approving or disapproving it."[52] This was essential, for the cultist label had not been removed and it was quite possible that the AMA house, if it had the chance, might very well veto the merger plan. Among the major provisions of the contract were, first, that the College of Osteopathic Physicians and Surgeons, which would change its name to the California College of Medicine, would offer to all of its living graduates and those DOs from other schools who held valid physician and surgeon licenses in the state a Doctor of Medicine (MD) degree. This would be an academic degree, the recognition of which for the purpose of licensure would depend upon the laws of the various states. However, in California, statutory provision would be made to accept it for all purposes connected with the practice of medicine. Second, those DOs who chose to accept this MD degree would thereafter cease to identify themselves as osteopathic practitioners in any manner. Third, the California College of Medicine would henceforth be a medical school affiliated with the Association of American Medical Colleges and end its teaching of osteopathy. Fourth, the CMA would absorb ex-DOs within the existing forty-county medical society structure, although during the transition period they would be segregated into a special forty-first society. Finally, the ex-DOs would support legislative action implementing the agreement, including the revision of the 1922 osteopathic initiative that gave them an independent board, to insure that there would be no new future licensing of DOs in the state. Those DOs already licensed in California who decided not to join the merger and retain their DO degrees would continue to come under the jurisdiction of the independent osteopathic licensure board until they numbered fewer than forty, at which point the board would be completely abolished and its activities taken over by the MD board.[53]

The CMA house on May 3 passed the agreement by a vote of 296 to 63. Two weeks later the COA house voted 100 to 10 to accept it. Later that month the Board of Trustees of COP&S were assured by representatives from the Association of American Medical Colleges that with some relatively minor organizational and staffing changes their institution would become a fully accredited medical school. The COA leadership also promised that the new Forty-First Medical Society would continue to financially support the institution. After a bitter debate, the Board of Trustees narrowly voted, 13 to 11, to go along with the conversion. On the 14th and 15th of July, some two thousand DOs, gathered in the auditorium of Los Angeles County General Hospital, received their new MD degrees.[54]

After these actions, the battle over the ballot initiative implementing the merger, played out during the next year, seemed anticlimactic. The measure was supported by the ex-DOs, the CMA, both houses of the legislature, the Democratic governor, Pat Brown, and his Republican challenger, Richard Nixon, as well as numerous civic organizations. The OPSC, which had been unable to stop any of the previous legal steps on the road to amalgamation as they moved through the courts, had to face this opposition alone. A pledge by the AOA to provide a sizable war chest had been withdrawn when it became clear that such a contribution was political and would remove the association's tax-exempt standing, since it, unlike the AMA, was registered as an educational organization with the Internal Revenue Service. With OPSC unable to generate sufficient capital to get its message across, their defeat was all but assured. A final count of the votes revealed 3,407,957, or 69 percent, marking their ballots "yes"; only 1,536,470 or 31 percent, registered "no." The merger had been completed.[55]

REAFFIRMATION & EXPANSION

To many outside observers the events taking place in California, from the initial public announcement of a merger plan to its implementation in 1962, seemed to signal the first step in the inevitable absorption of the DOs as a group into the allopathic medical profession. Whatever agreements had to be worked out, complete countrywide amalgamation was viewed as a foregone conclusion. There was no possibility that the MDs would continue to let the DOs remain independent or that the AOA could continue to resist the pull of the AMA upon its members. Movements such as osteopathy, homeopathy, and eclecticism, it is generally believed, have a natural life cycle. They are conceived by a crisis in medical care; their youth is marked by a broadening of their ideas; and their decline occurs when whatever distinctive notions they have as to patient management are allowed to wither. At this point, no longer having a compelling raison d'être, they die. While examples of this pattern are not difficult to find in the United States and elsewhere, this type of explanation tends to downplay, if not ignore, specific highly individualized historical conditions. Whether osteopathy would be able to survive the California merger intact or would go the way of homeopathy and eclecticism would depend not upon some deductively arrived at natural law but upon actual social circumstances over time.

CALIFORNIA AFTERMATH

Throughout the 1950s it was readily apparent that the AMA could not forge any coherent national policy with respect to the DOs. Indeed, if the AMA executive secretary had not ruled that the amalgamation being arranged in California was a local matter and therefore not subject to action by the House of Delegates, it is entirely possible that the final agreement would not have been allowed. In succeeding years this lack of uniformity and consistency on the part of the AMA and other significant groups within organized medicine would severely hamper their efforts to obtain what they would regard as a satisfactory conclusion to "the osteopathic problem."

In June 1961, while the California merger was still in process, the AMA Judicial Council delivered a special report on the association's position on voluntary relations between MDs and DOs. Noting that "there cannot be two distinct sciences of medicine or two different yet equally valid systems of medical practice," it declared that the changes occurring within osteopathy indicated the desire by a significant number of DOs to give their patients scientific medical care. Because of this, the council said, "Policy should now be applied individually at [the] local level according to the facts as they exist. The test should now be: does the individual doctor of osteopathy practice osteopathy or does he in fact practice a method of healing founded on the scientific basis? If he practices osteopathy, he practices a cult system of healing, and all voluntary professional associations with him are unethical. If he bases his practice on the same scientific principles as those adhered to by members of the American Medical Association, voluntary professional relations with him should be deemed ethical."[1]

Many AMA house delegates thought this proposed policy was far too liberal, since it would serve to make the issue of ethical relations subject to the discretion of individual MDs. A number of component societies, particularly those whose states restricted practice rights for DOs, continued to be bitterly opposed to any interprofessional contact and thus were unwilling to support the report as it stood. As a compromise, it was amended to give each state society the right to make the determination as to whether or not its members could voluntarily associate with osteopathic practitioners on a professional basis. In this form, with almost the exact wording of the 1955 Cline committee recommendation, the policy was approved.[2] The Judicial Council also urged that local liaison committees, if not already in existence, be established to conduct talks with their DO counterparts, and this was fully accepted. With regard to some states the report noted, "It might be possible to initiate and complete negotiations such as have been and are being carried out in California."[3] To assist the state societies in the formulation and carrying out of their plans, the AMA board created the Committee on Osteopathy and Medicine the following year.

While the AMA was characterized by sharp division within its official ranks over the question of MD-DO relations, the AOA, for its part, was united, at both the national and state levels. At its July 1961 annual meeting, held in Chicago, the AOA delegates in strong language reaffirmed the so-called "Michigan Resolution" of 1959, which declared their intent to maintain the status of osteopathy as a separate and complete school of medicine. Coupled with this was a sharply worded statement by the AOA Board of Trustees attacking the premises of the recent report of the Judicial Council, noting:

> It may be true that there cannot be two sciences of medicine, but the AMA fails to recognize that while medicine employs scientific knowledge, the *practice* of medicine is not science per se. It is unrealistic to hold that the practice of medicine is pure science. It is equally unrealistic to insist that only one system of medical practice, that system officially approved by a political body, can be valid. . . . The AMA holds that if an individual doctor of osteopathy practices osteopathy he is a cultist and all voluntary professional associations with him are unethical. However, if he bases his practice on the same scientific principles as those adhered to by members of the AMA voluntary relationships are ethical. This policy has two fallacies: First, it assumes that the osteopathic concepts are diametrically opposed to accepted scientific fact and that osteopathic physicians do not employ accepted scientific principles in their practice. Second, it condemns a system of practice without understanding or defining it, or, in fact, defining what is accepted scientific medical practice.[4]

As for the mechanisms of declaring whether relations between MDs and DOs were ethical, the AOA house stated that the AMA, in granting their state societies the right to decide who shall and shall not be legitimate, hoped that organized osteopathy would be weakened from within and that through internal dissension it would be eliminated. "The osteopathic profession," the AOA House of Delegates declared, "will continue as a separate and complete school of medical practice and . . . it will resist all efforts to be absorbed, amalgamated or destroyed, be it through overt political maneuvering, or through the guise of making its individual members conform to the scientific dictates of the AMA."[5]

In dealing with liaison committees established by state medical associations, the AOA divisional societies adopted a common strategy. When approached, osteopathic representatives would announce that before any other interprofessional matter could be brought up, all medical society restrictions on voluntary MD-DO relations had to be removed. In those cases where this was done, the DOs then agreed to discuss mutual problems, though they turned a deaf ear to the subject of amalgamation. By 1965 only fourteen state medical associations had approved voluntary interprofessional relations. In some states the MDs held the approval of voluntary professional association hostage to merger negotiations; in others, MD antipathy towards the DOs precluded any formal discussions whatsoever.[6]

While organized osteopathy was united against amalgamation, there was some dissension within the rank and file. However, unlike in California, where absorption proponents were an active force within their state society, controlled political offices, and were thus able to maneuver the COA towards their desired goal, in other states DOs who supported the merger concept were more likely not to be members of the AOA or their divisional society; and if they were members, they had not attained positions of influence. Opposition to the AOA policy line, therefore, was mostly scattered, unorganized, and lacking in effective leadership. The one exception occurred in the state of Washington, where in 1962 a faction of dissident DOs broke away from the official state osteopathic association, formed their own group, and sought to arrange a merger between themselves and the state medical society. With the support of the latter, they founded a "paper college" to award MD degrees, valid for the purposes of licensure, to "qualified" Washington DOs. Nothing came of this plan, however, as the Washington State Supreme Court, in a unanimous ruling, declared that the Washington Board of Medical Examiners' decision to approve the paper college as a medical school was "subterfuge, was palpably arbitrary and capricious, and was void in all respects."[7]

A significant number of DOs around the country were undecided about merger. Troubled by their own special problems, whether it be poor public perception, denial of staff privileges at local hospitals, or ineligibility for participation in insurance plans, they harbored genuine doubts as to the wisest position for them to take. To help counter such wavering, the AOA organized and conducted a series of "regional town meetings" across the country in which officials explained recently adopted policy positions as well as the association's efforts on the legislative and judicial fronts to break down discriminatory barriers.[8] At such gatherings AOA leaders addressed the fears and frustrations of concerned DOs, and they convinced many who attended of the "rightness" of the association's stand. Nevertheless, other DOs adopted a wait-and-see attitude, letting time pass to enable them to determine for themselves how well the California plan was working out before coming to any hard-and-fast conclusions.

In the years following the California merger it became evident to the ex-DOs who participated that their move had both positive and negative features. On the favorable side, the vast majority seemed to be quite happy with the new MD initials behind their name. Although one AOA leader warned them that they and members of their family would be forever subjected to the whisper, "He is an osteopath who was given an MD degree," this did not appear to be a significant problem.[9] Patients readily accepted the changed designation as well as the new diploma on the wall, and most ex-DOs felt relieved at no longer having to answer such questions as, "What kind of doctor are you anyway?"[10] The most satisfied of the ex-DOs were clearly the general practitioners. Aside from their new degrees, they found that they could now obtain admitting privileges at hospitals that had once barred them, that their malpractice rates as a result of joining the CMA were substantially lower than they had been previously, and that they could freely consult with a wider range of specialists.[11]

Nevertheless, major problems did surface. As part of the amalgamation contract, all ex-DOs were to be temporarily segregated in a special forty-first component society of the CMA until they could be fully integrated into the other forty county societies. Though most of the ex-DOs had no difficulty in being assimilated, a significant number did. As late as 1967, five years after amalgamation, approximately 10 percent of the original group had not been granted regular local membership. This proved to be a source of some embarrassment and discontent.[12]

Far more serious was the status of ex-DO specialists—a subject that was left unresolved in the merger agreement. Under existing AMA requirements, all candidates for specialty board certification had to graduate from an accredited medical school and receive their postgraduate training in an AMA-approved hospital program. This meant that ex-DOs, including those who had been certified by an AOA board, could not receive any consideration for such certification. Though CMA officials reportedly pledged to work for changes in allopathic specialty board policies, no movement in that direction was forthcoming. What the CMA agreed to do was to inspect the DO specialists' osteopathic credentials and then issue a certificate stating that they were found to be in order. While this document may have been suitable for hanging in the office to impress one's clients, it could not help the practitioner gain staff privileges at other than formerly osteopathic hospitals.[13] Another consequence of this lack of proper certification was a decline in the number of patients regularly referred to these specialists. Before the merger, DO generalists would be more likely to send a patient an inconvenient distance to see a DO gynecologist, internist, or surgeon. But now, as the ex-DO general practitioners made new professional acquaintances, they began to refer patients to specialists more on the basis of their credentials and proximity. Furthermore, most of those practitioners whom the ex-DOs called "congenital MDs," that is, those who graduated from an AMA-accredited school, were unwilling to send patients to "acquired MDs."[14]

In the years following the merger agreement, a number of formerly osteopathic hospitals in California found themselves in financial difficulty. Some institutions reported a staff loss of up to 20 percent, as local DO general practitioners began affiliating with MD facilities. In some cases, the loss was made up for by the addition of traditional MD's to the staff, but when the AOA surveyed the hospitals in 1965 most reported that their occupancy rate was lower than it had been before the merger. A few institutions saw themselves as eventually having to go out of business, while others anticipated that they would become satellite facilities for major MD hospitals.[15] Also, all ex-DO facilities lost their intern and residency programs, because, with the exception of the new Los Angeles County Hospital, they were simply too small to qualify for AMA approval. Therefore, these formerly osteopathic inpatient facilities were no longer teaching oriented, a change not a few of the staff regretted.

What was once the private College of Osteopathic Physicians and Surgeons and is now the California College of Medicine became state-supported in 1964; a new campus was established for it at the University of California at Irvine several years later. With a far greater source of revenue at its disposal, new equipment was purchased, and many more full-time instructors were hired. As it was now an orthodox medical institution, it could become affiliated with a number of large MD hospitals, thus improving training opportunities. Losing out in this process were many part-time and voluntary ex-DO faculty members whose services were no longer required. Also affected were a number of full-time ex-DOs who, while not removed from the staff, found themselves maneuvered out of positions of authority in favor of congenital MDs.[16]

A final critical problem arising from the merger was the validity of the acquired MD degree as a basis for licensure everywhere outside California. By 1966, courts in ten states had ruled in favor of those examining boards which rejected applications from holders of the 1961 diploma on the grounds that theirs was an academic, not a professional, degree. Only those California College of Medicine students completing their training in 1962 or later were considered graduates of an AMA-accredited institution.[17]

The two major AOA publications, the *JAOA* and *The D.O.*, continuously pointed out and amplified all these problems to their readers as evidence that amalgamation was a failure. Editorials blasting the holders of what was labeled "the little md" were occasionally coupled with letters from ex-DOs who voiced deep disappointment with all or certain features of the merger. Although the picture drawn by the AOA was one-sided, it was nevertheless apparent from a reading of even generally pro-merger articles in nonosteopathic journals that not everything had worked out as well as all ex-DOs had hoped.[18] These perceived inequities and difficulties only served to make many undecided DOs around the country wary of amalgamation—or at least amalgamation of the California variety. In a 1972 mail survey of DOs in twelve geographically scattered states conducted by the independent journal *Osteopathic Physician*, only 17 (or 7.8 percent) of 218 practitioners responding answered "yes" to the question, "Do you view the merger in California as a satisfactory one?"[19]

In addition to the California situation, a major factor that would lead undecided DOs to shy away from supporting merger was the perceived continued inability on the part of the AMA and other medical groups to treat them with what they believed was sufficient professional respect. Indeed, where organized medicine altered existing discriminatory policies towards the DOs, its only motivation seemed to be the desire to solve the problem these policies caused for MDs—not to eliminate "gross injustices" against osteopathic practitioners. For example, in 1959 the American Hospital Association (AHA) decided to change its longstanding policy barring joint- or mixed-staff institutions from membership. However, this reversal occurred only after the association became the focus of intense pressure from public hospitals that were being forced by court or legislative action to allow DOs access to their institutions and to give them staff appointments. Under the AHA's revised rules, DOs could become staff members, but general supervision of the clinical work was to remain the responsibility of MDs alone.[20] American Hospital Association membership was a necessary prerequisite for eligibility for accreditation by the Joint Commission on the Accreditation of Hospitals. In 1960, again only under strong pressure from public hospitals, the Joint Commission made the appropriate adjustments to permit these AHA-member mixed-staff institutions to be inspected and accredited.[21]

To many DOs the "true" attitude of the AMA and other allopathic medical groups towards the osteopathic profession could be seen in a variety of policy decisions. When it came to supporting opportunities and responsibilities for osteopathic physicians and surgeons equal to those enjoyed by MDs in public hospitals, organized medicine said no; when it came to changing practice laws that discriminated against DOs, organized medicine was generally opposed; finally, when it came to pending federal legislation to underwrite the expenses of health profession schools, the AMA would testify that osteopathic institutions should be excluded.[22] Furthermore, many of those DOs who read *JAMA* articles pertaining to osteopathic medicine, either in the original or as reprinted elsewhere, felt denigrated or insulted by their assumptions and tone. Particularly galling was the fact that in article after article, DOs were referred to as "osteopaths" in contradistinction to "physicians"—a title used to denote MDs only. Based on the actions and rhetoric of the AMA, most DOs came to the conclusion that the association was unwilling, or perhaps incapable, of dealing with them as equals. Rather, it appeared that organized medicine regarded the osteopathic profession as nothing more than a nuisance which had to be eliminated one way or another.

THE NEW AMA OFFENSIVE

By the mid-1960s it was apparent to the AMA leadership that the osteopathic profession was standing firm. The policy of allowing the state medical societies to decide the cultism issue had not served to bring amalgamation closer to fruition. Furthermore, the unresolved problems of the California merger had caused considerable skepticism among individual DOs. However, most distressing to the AMA leadership had to be the fact that the merger itself was having the unintended consequence of permitting the osteopathic profession to make key political gains, thus serving to increase the strength of the AOA.

National and state osteopathic societies used the facts that DOs in California had become MDs without any additional educational requirements and that COP&S was so quickly accredited as an MD-granting institution to press their case for revision of discriminatory policies against DOs. They argued convincingly that what the merger had shown was that whatever gaps there might be in the quality of training between DOs and MDs, they were no longer significant. As a consequence, legislatures in several limited licensure states subsequently changed their laws to make

DOs eligible for the same scope of license available to MDs. The merger had a similar impact at the federal level. In 1963 the U.S. Civil Service Commission, specifically citing events in California, announced that for its purposes the MD and DO degrees were henceforth to be considered equivalent. In 1966 Secretary of Defense Robert McNamara, using legislative authority granted to his office a decade earlier, ordered all the armed services to accept qualified DOs as military physicians and surgeons for the first time. Also that same year, the AOA won a major victory when it was accepted as an accrediting agency over osteopathic hospitals for the purpose of determining an institution's eligibility for participation in the Medicare program (Public Law 89-97, July 30, 1965). Thus, the DOs were increasingly obtaining on their own some of the benefits the MDs could offer through amalgamation.[23]

In response to these important gains for osteopathic physicians, the AMA in the late 1960s adopted a series of new resolutions aimed at destroying the AOA. Aiming to take away many of its colleges, students, interns, and residents, as well as a large proportion of its members, the AMA sought to quickly force the issue of absorption. The first actions came in July 1967, when the AMA House of Delegates authorized its Board of Trustees to begin negotiations promptly with all the DO schools for the sole purpose of converting them to orthodox medical institutions. In order to place pressure upon the colleges to bargain, the AMA house authorized the Council on Medical Education "to establish means by which selected students with proven satisfactory scholastic ability in schools of osteopathy may be considered by schools of medicine for transfer into medical school classes." In short, the colleges were warned that if they resisted the AMA overture, they would soon find themselves with a sharply depleted enrollment. In adopting these actions, the AMA house noted, "The primary issue at the present time in the relationship of medicine and osteopathy seems to be not that of cultism as opposed to science. Rather the issue appears to be one level of medical education and practice to another and lower level of medical education and practice."[24]

The next month, the AOA House of Delegates adopted a resolution that declared in part, "The AMA contention that osteopathic education needs to be improved is obviously not shared by recognized educational accrediting agencies, by state licensing bodies or by the millions of Americans who prefer osteopathic care. . . . The AMA stands alone in its assessment of osteopathic education, but the osteopathic profession stands together in vigorously opposing this arrogant policy of academic piracy."[25]

Indeed, the colleges did hold together, although a difficult economic situation at the Des Moines school led its administration to hold talks with AMA and Association of American Medical Colleges representatives. However, this threat to solidarity was eliminated when those college officials resigned under AOA pressure and were replaced by individuals who opposed merger.[26]

Disappointed with an initial lack of movement, the AMA soon followed with two other major policy shifts. In December 1968, the House of Delegates passed resolutions that, first, encouraged each county and state medical society to change its by-laws so that it might "accept qualified osteopaths as active members," and, second, urged that each of the boards of medical specialties change its rules in order to "accept for examination for certification those osteopaths who have completed AMA-approved internship and residency programs and have met the other regular requirements applicable to all board candidates." As specialty boards declared their intent to permit examination of DOs, appropriate AMA-approved residency programs would be opened to qualified osteopathic graduates. Determination of qualification for acceptance into a given program would be left up to the medical staff of the hospital or to the county medical society.[27] In June 1969, the AMA house extended to DOs membership in the national association and officially changed the "Essentials of Approved Residencies," clearing the path for acceptance of DOs into those programs in which the respective specialty boards had agreed to examine them for the purposes of certification. At that time, five boards—pathology, pediatrics, physical medicine and rehabilitation, preventative medicine, and radiology—had done so. By 1971 the number had increased to thirteen.[28]

These actions posed serious potential problems for the AOA. As early as 1968 the issue of belonging to an allopathic medical association arose in connection with two DOs who had accepted associate membership status in the Michigan State Medical Society. The AOA House of Delegates reacted by adopting a resolution declaring that "any member accepting associate membership in the American Medical Association or any of its political divisions is acting contrary to the best interests of the American Osteopathic Association and shall be subject to discipline up to and including expulsion." The following July the AOA House of Delegates clarified this resolution by interpreting "political divisions" to mean national, state, divisional, or county medical societies.[29]

In 1971 both the Iowa and the Pennsylvania osteopathic delegations offered resolutions to the AOA house seeking to reverse this policy. In each of these states, particularly in areas where there were no osteopathic hospitals, DOs found themselves removed from or denied staff privileges at local public and private facilities. This action was not due to their osteopathic identity but because these institutions, as a result of the AMA policy shift, now insisted that all DOs, like their MD physicians, be members of the county medical society. Osteopathic practitioners in these states argued that they had no realistic choice but to comply. During the floor debate in the AOA house, delegates from other state osteopathic associations responded that they had encountered the same allopathic hospital maneuver but had overcome the problem by seeking and receiving legislative and judicial relief. When the Iowa delegates admitted that they had not exhausted their legal options, much of whatever sentiment there was for their measure evaporated. Pennsylvania thereupon withdrew its proposal, and the 1968 policy as amended in 1969 was reaffirmed. Another effort at overturning the established rule was made in 1973, but it met a similar fate.[30] At the end of 1978 only 417 osteopathic practitioners (2.4 percent of all listed DOs) had joined the national AMA.[31]

The issue of postdoctoral opportunities for DOs in allopathic hospitals presented a far more complex and difficult situation for the AOA. While the association had made considerable strides in upgrading its standards regarding internships and residencies in recent decades, serious weaknesses remained, particularly in some of the specialties. Since the DO hospitals utilized for such training were typically smaller than those of their MD counterparts, the range and depth of experience offered was not always comparable. Also, in some established fields like dermatology and proctology, there were no hospital residencies, only preceptorships; and in other specialties, such as psychiatry, few programs existed. Finally, there was the question of those DOs now entering the armed forces and Public Health Service. The only manner in which they could receive formal postdoctoral training while on duty was in federal hospital programs accredited by the AMA.

When the matter of postdoctoral training first came up before the AOA house in July 1969, no definitive action was taken. Instead, it was decided to give the AOA Committee on Post-Doctoral Training the authorization to provide applications for nonosteopathic hospital intern and residency programs on an individual basis.[32] This absence of a clear policy led to considerable confusion among the ranks, which was only partly relieved at a joint conference between the AOA board and Associated Colleges representatives held that December. Following this meeting, AOA president J. Scott Heatherington, DO, addressed a letter to all osteopathic students, faculty, and administrators in which he stated the association's position. "The AOA," he wrote, "recognizes that there are a few highly technical subspecialty fields in which neither the osteopathic nor allopathic approach to health care can be clearly differentiated at this time. Within these limited fields there may be legitimate grounds which enable osteopathic physicians to participate in training under allopathic auspices, but only so long as such sub-specialty training clearly augments, not replaces osteopathic training in the major specialty fields."[33] Students were warned that before they could receive the AOA's blessing to enter an AMA residency, they had first to complete an AOA-approved rotating internship—that is, one in an osteopathic hospital recognized for that purpose or a federal hospital, "as long as it fits the rules." The next step was for the student to provide the AOA with a detailed outline of the AMA residency training program into which he or she had been accepted.

At its July 1970, meeting, the AOA house gave its approval to this basic plan, though again specific criteria under which a candidate might or might not be allowed to take an allopathic residency awaited formulation.[34] In the case of some specialties, such as general surgery and internal medicine, for which there was a sufficient number of residency programs in osteopathic institutions to meet the needs of DO postgraduates, both the AOA and the hospitals feared that these might be bypassed by osteopathic trainees unless further restrictions upon allopathic appointments were established. Consequently, in 1970 and 1971, the respective specialty boards in these and other fields began to change their certification requirements to insist that one or more years had to be spent in an osteopathic hospital residency before a student could be given credit for nonfederal allopathic training. Finally, after much delay, AOA policy had taken form.

The optimistic prediction within organized medicine that there would be an immediate mass defection of DOs from AOA-approved postdoctoral programs was not fulfilled, although in the first few years the number of new osteopathic physicians entering nonmilitary allopathic programs upon graduation was certainly significant. According to AOA-released data, 12 percent of the class of 1970 followed this route.[35] Data subsequently collected suggest that this figure remained stable through 1973. Afterwards the total began to drop: 9 percent in 1974, 3 percent in 1975 and 1976—this despite the fact that more allopathic hospitals were opening up their programs to DOs. In the total number of osteopathic physicians training in allopathic hospital programs, including those approved by the AOA, a similar pattern may be seen. The figure rose rapidly each year, peaking at 608 in 1973, but declined to 449 in 1977. Meanwhile, the number of residents in osteopathic hospitals made a modest gain between the 1972–73 and 1976–77 contract periods.[36]

Three principal reasons may be offered as to why the large break anticipated by the AMA did not occur. First, most DO students and recent graduates perceived that their postgraduate programs were, by and large, satisfactory and that the training they would receive was comparable to that available in an allopathic hospital. Second, some prospective trainees believed they would be looked down upon or discriminated against in an MD environment. And third, some who wanted to enter an allopathic program were fearful of possible disciplinary actions by the AOA should they not follow its guidelines. One reason a decline in osteopathic participation in non–AOA-approved programs occurred after 1973 appears to be a landmark Arizona Court of Appeals decision handed down that year concerning a DO with strictly allopathic postdoctoral credentials who had been denied a medical license by the state board of osteopathic examiners on the grounds that he had not served a one-year rotating internship in an AOA hospital program as required by law. The DO, backed by the AMA, brought suit, claiming that training under allopathic auspices was equivalent and thus should be accepted. The court, however, turned aside this argument and upheld the board's decision. Since at least thirteen other state boards, including the osteopathic strongholds of Michigan, Pennsylvania, Florida, and Oklahoma, were covered by similarly worded statutes, some students who had planned to bypass the AOA-approved routes undoubtedly thought better of the idea.[37]

THE COLLEGE BOOM

One of the justifications given by the California delegation for the decision to merge with the CMA was that the osteopathic profession was not growing. As a result, the prospects for its becoming socially visible were not good. Indeed, if one surveys the number of graduates produced by the colleges each year prior to 1962, no pattern of continuous expansion can be discerned, only ups and downs related to entrance requirements, economic conditions, and war. What gains there were in the total number of listed DOs during this period were simply a reflection of the fact that as an occupational group, osteopathic physicians were getting older. Now, with one less college, some two thousand fewer practitioners, and a loss of between ninety and one hundred new graduates each year, leaders within the osteopathic profession saw the necessity not only of replenishing its ranks but of going well beyond its premerger totals of schools and practitioners.

In their struggle to increase their numbers, DOs were aided by outside factors. Throughout the 1950s, claims were being made that there either was, or soon would be, a serious shortage of practicing physicians in the United States since the number of medical schools and graduates was not keeping pace with the postwar growth in population. The issuance of two Department of Health, Education, and Welfare studies, the Bayne-Jones (1958) and Bane (1959) reports, lent weight to these conclusions, and attention soon shifted to what the federal government could do to eliminate the perceived problem. This, along with concern about the overall quality of medical training, led to the passage of the Health Professions Education Act of 1963 (Public Law 88-1929), which authorized a program of matching federal funds for construction and improvement of medical schools, together with a program of making loans to students in medicine, osteopathy, and dentistry. An amendment to this act two years later (Public Law 89-290) established a scholarship program, and all of the aforementioned provisions were later included in the Health Manpower Act of 1968 (Public Law 90-490). Federal aid to osteopathic as well as other professional schools would be further increased with the signing into law of the Comprehensive Health Manpower Training Act of 1971 (Public Law 92-157), which raised support levels for construction, replaced institutional grants with capitation grants to stimulate further enrollment gains, authorized special project moneys, and broadened student loan provisions.[38] From fiscal year 1965 through 1976, the Chicago, Des Moines, Kansas City, Kirksville, and Philadelphia schools received a combined total of $65.8 million through these specific programs.[39]

Other new sources of funding were made available. The legislatures of Pennsylvania (1966), Illinois (1970), and Iowa (1973) passed bills inaugurating ongoing educational assistance programs to their respective colleges of osteopathic medicine, in addition to authorizing separate grants for new construction. For the first time, assistance was secured from major philanthropic foundations, as well as from the pharmaceutical houses. Increased support from traditional sources also helped. Between 1961 and 1975, the Osteopathic Progress Fund, supported by DOs in the field, channeled slightly over $16 million into the schools, and the colleges themselves roughly tripled their tuition. In a federally sponsored study published in 1974, it was found that, while the median spending level of sampled DO schools was still lower than that of sampled MD institutions, all osteopathic colleges examined were now within the total range of MD schools studied with respect to the amount of money each spent per student for educational purposes.[40]

Several significant improvements were made in these five colleges between the time of the California merger in 1962 and the late 1970s. First, the qualifications of their students steadily rose. During the 1958–59 academic year, 72 percent of entrants held bachelor's or advanced degrees. By 1968–69, this had climbed to 88 percent, and during 1978–79 the total exceeded 95 percent.[41] Second, more faculty members, particularly full-time staff, were hired.[42] Finally, equipment and facilities were improved. The Chicago College added two new wings to its existing hospital (1963–70), built a new basic science building (1968), opened a new $12.3 million outpatient clinic (1978), and completed construction of an $18 million, 200-bed satellite facility (1978). The Kansas City College added a new library (1968), lecture halls (1971), and a $29 million, 426-bed teaching hospital (1972). The Philadelphia College built a new campus that included a 250-bed facility (1968), the Kirksville school added a new research building (1963) and completed a major addition to its hospital (1971), and Des Moines moved its campus to more spacious quarters (1972). One indirect measure of improved standards and conditions within these schools was the overall performance of DO candidates before MD and composite licensure boards. Data published by *JAMA*, suggested that by the early 1970s there were no significant statistical differences between DOs and U.S.-trained MDs in passing such examinations.

Even more important to the future of the profession, the perceived overall shortage of physicians helped spur the establishment of new osteopathic schools, particularly as the existing DO colleges had a proven record of producing a high percentage of the type of doctor most in need, that is, general practitioners who were most likely to locate in rural and inner city areas.

The first and most significant battle to establish a new college occurred in Michigan—which, after the California merger, now had the largest number of osteopathic physicians in any state. Although hampered by the lack of a school, many DOs were drawn to practice within Michigan by an attractive licensure law and public acceptance. The limited licensure of

Table 4. Examination Results before Medical and Composite Licensing Boards
for U.S.-Trained MD and DO Physicians and Foreign Medical Graduates,
1955–59 through 1970–72

Period	U.S. MD Examinees			U.S. DO Examinees			Foreign Medical Graduates		
	Examined	Passed	%	Examined	Passed	%	Examined	Passed	%
1955–59	30,184	28,903	95.7	1,174	954	81.2	11,192	6,787	60.6
1960–64	25,995	25,360	97.6	1,980	1,678	84.7	14,534	9,959	68.5
1965–69	23,364	22,321	95.5	2,135	1,887	88.4	20,800	13,242	63.7
1970–72*	15,922	14,368	90.2	1,401	1,241	88.6	25,725	16,477	64.1

Source: "Medical Licensure Statistics," *Journal of the American Medical Association* 161 (1956): 341; 164 (1957): 426; 167 (1958) 594; 170 (1959): 573; 173 (1960): 387; 176 (1961): 701; 180 (1962): 847; 184 (1963): 788; 188 (1964): 880; 192 (1965): 858; 196 (1966); 861; 200 (1967): 1058; 204 (1968): 1070; 208 (1969): 2086; 212 (1970): 1875; 216 (1971): 1786; 220 (1972): 1607; 225 (1973): 301.
*Data were not published after 1973. Accuracy of data covering 1973 is in dispute.

its neighbor Illinois discouraged graduates of the Chicago school from staying within that state. However, after Illinois joined the unlimited licensure ranks, in 1955, and COP&S closed, in 1961, a number of influential Michigan DOs believed they would have to establish their own college if they were to maintain or increase their ranks.

In May 1963, the Michigan Association of Osteopathic Physicians and Surgeons (MAOP&S) House of Delegates unanimously threw its support behind plans for a new school, which was to be established near East Lansing, home of Michigan State University (MSU). However, when MSU announced shortly thereafter that it was in the process of developing an MD-granting institution, the osteopathic college committee shifted the location to Pontiac. In March 1965, a charter was obtained and architects were hired to design the campus.[43]

Meanwhile, representatives and other advocates of the proposed college began lobbying for state aid. They referred legislators to recent surveys conducted by a commission appointed by the governor showing a need for yet another medical school since Michigan ranked only twenty-fifth among all states in physician-population ratio.[44] As to why this should be an osteopathic rather than an allopathic school, the DOs pointed out that as most of them were general practitioners, disproportionately located in underserved areas, they were filling the health care gaps that the MDs had created. Thus, to solve the perceived physician workforce problem, it made more sense to invest in osteopathic medical education. These arguments interested the legislature, which in June 1965, passed a capital outlay bill providing money for a feasibility study. That same month the MAOP&S house assessed each member of the association $2,000 payable over the next ten years to raise $3 million for the institution, and unveiled plans to amass $5 million elsewhere so as to qualify under the Health Professions Education Act for another $16 million in its two-to-one matching program.[45]

In October 1966, the Michigan Senate by a vote of 22 to 7 passed a bill creating the authority for the establishment of a state-supported osteopathic school. Not unexpectedly, the Michigan State Medical Society protested. During hearings on the measure before the House State Affairs Committee the next month, the Medical Society's president appeared, forcefully arguing that amalgamation between the two professions was imminent. A state-financed school "just for osteopaths," he maintained, would be absurd, since at least 75 percent of all Michigan DOs favored merger. This assertion was vigorously rebutted by MAOP&S representatives. With no concrete data available, the House State Affairs Committee could not determine the accuracy of either contention, so it decided to commission a confidential mail ballot addressed to all DOs and MDs practicing in Michigan to measure their opinions. The results, released in early 1967, were unambiguous. To the question "Do you believe amalgamation of allopathy and osteopathy would be in the best interest of the state?" 87.3 percent of the DOs who responded said "no." On the question "Should the state give support to the osteopathic school?" 93.3 percent of the DOs answered "yes." Results from the MDs polled revealed opposite responses in approximately the same proportions. With this new information, the house committee voted 10 to 1 in favor of the College Authority.[46]

The Michigan State Medical Society however, did not give up. When the measure came before the full house for consideration in mid-1967, it lobbied intensively and successfully for the bill's defeat, which was by a margin of only two votes. The legislature, though, had not closed the door on the project, having already allocated another $50,000 for further study and development. The following year it appropriated $75,000 more. Finally, in 1969 the question of state support came before the legislature once again. This time the osteopathic forces were much better prepared. They responded well to the objections raised by the medical opposition and helped push their bill through both houses and secure the governor's signature.[47]

Under the new statute, the osteopathic college would become an integral part of one of the three existing state universities. Further details were to be decided by the Michigan Board of Education and agreed to by the board of trustees of that institution. After involved negotiations, Michigan State University was chosen and accepted. Meanwhile, the board of trustees of the proposed school had previously voted to press ahead with or without state aid. It had already begun its first class in Pontiac in the fall of 1969 and would start a second year there before the whole campus would be transferred to East Lansing, where existing buildings were being remodeled for its use. The new Michigan State University College of Osteopathic Medicine (MSU-COM) would share some facilities with the MD-granting school that had been created on campus—the College of Human Medicine—but each would be governed by a separate budget and administration. While each college would use the same pool of basic science faculty, some classes for DO and MD students would be held separately.[48]

The establishment of MSU-COM was significant in at least three major respects. It was the first new school of osteopathic medicine to have been founded in several decades and helped to show that the profession was not content with merely maintaining its existing number of colleges and graduates. Second, it was the first university-based osteopathic school, thus allowing the profession to achieve greater status in the academic community. Third, having a DO and an MD college existing side-by-side on the same campus gave visible expression to the contention by AOA leaders that the two medical professions were "separate but equal."

At the same time that Michigan DOs were making plans for their new school, osteopathic practitioners in Texas were working towards the same end. Enrolling its first class in the fall of 1970, the Texas College of Osteopathic Medicine (TCOM) began inauspiciously, housed initially on the top two floors of Fort Worth Osteopathic Hospital. However, the following year more suitable facilities for basic science instruction were obtained and utilized. Although founded as a private institution, TCOM began receiving some state aid in 1971, and the next year it signed a contract with North Texas State University (NTSU) in Denton for the use of classrooms, faculty, laboratories, and offices. In 1973 state assistance was significantly increased with the passage of an appropriations bill providing TCOM with capitation funds—$11,625 for each bona fide Texas resident enrolled. Two years later a formal agreement was negotiated and signed under which TCOM would become a public institution under the control of the Board of Regents of NTSU. Thus, the profession had its second university-affiliated medical school.[49]

Given the successful efforts of Michigan and Texas, DO groups in other states began pushing in earnest for their own institutions. The next to be established were the Oklahoma College of Osteopathic Medicine and Surgery in Tulsa and the West Virginia School of Osteopathic Medicine in Greenbriar, both opening in 1974. In the case of Oklahoma, the legislature was impressed by what DOs were already accomplishing in the state—providing medical services in high-need areas—and was therefore willing to expand their role by creating a freestanding public college.[50] In West Virginia, on the other hand, DOs had made comparatively little impact on health care delivery, since there were only about seventy active practitioners. Nevertheless, a determined West Virginia Osteopathic Society, recognizing the dire need for more physicians in the Appalachian region, decided that they were best able to fill the gap. It purchased and remodeled a former military academy and began operations on a limited budget, backed by the necessary, although reluctant, AOA approval and the support of federal agencies that saw the school as an important experiment in increasing the physician workforce in economically depressed areas. The West Virginia legislature soon agreed, and the following year the institution was converted from a private to a freestanding public college.[51]

The drive to create more schools continued. In 1975 the Ohio legislature passed a bill authorizing the establishment of a state osteopathic school at Ohio University, which immediately transformed existing dormitories into offices, classrooms, and laboratories, enabling the school to accept its first class the following year.[52] In 1977 two more colleges began operation: the New Jersey School of Osteopathic Medicine, a state institution, part of what became known as the University of Medicine and Dentistry of New Jersey; and the New York College of Osteopathic Medicine, a private school affiliated with the New York Institute of Technology.[53] In 1978 another two private schools were established: the New England College of Osteopathic Medicine in Biddeford, Maine, a component of what became known as University of New England; and the College of Osteopathic Medicine of the Pacific (now a component of Western University), based in Pomona, California. This last school was made possible in part by a 1974 California State Supreme Court ruling that overturned the section of the merger legislation that barred any new osteopathic licensing in California.[54] Buoyed up by their success, California DOs who had remained loyal to the profession vowed to multiply their small numbers quickly and once again make osteopathic medicine a significant part of the health resources of the state.[55]

Between 1968 and 1980 the number of osteopathic schools rose from five to fourteen—an incredible leap in so short a period. These new institutions, along with increases in enrollment at already established osteopathic colleges, put the number of students far beyond the premerger average. In 1960 there were 1,994 students; by 1980 this number stood at 4,940. In 1960 there were 427 graduates, in 1980 there were 1,151. The reduction in number of osteopathic physicians nationally caused by the loss of the ex-DOs in California was quickly made up. The AOA directory premerger figure of 14,000 in 1961 was reached and surpassed in 1973. As of 1980 there were more than 18,000 listed DOs, and one study projected that there would be approximately 30,000 active osteopathic physicians and surgeons by 1990.[56]

To most DOs across the country, particularly those who had been in practice at the time of the California merger, all of this growth produced a psychological lift. The college boom, along with the success of their profession in withstanding AMA pressure for amalgamation, demonstrated to them that osteopathic medicine was not on the wane. Indeed, despite their growing concern with external economic and political forces increasingly impacting the entire health care industry, most DOs believed that the osteopathic profession was entering the most fruitful period of its history.

IN A SEA OF CHANGE

Having successfully resisted the AMA's aggressive efforts to achieve a national amalgamation of MDs and DOs, the osteopathic profession now faced quite a different threat to its autonomy. Government and private health insurers were in the process of transforming the entire health care system in response to significant and unrestrained annual increases in the cost of providing health care. In the 1970s, leaders of the osteopathic profession did not anticipate the vast power these "third parties" would eventually wield, nor did they imagine the impact they would soon have on health services in general or on osteopathic medicine in particular. Not until the 1980s did the osteopathic profession recognize that it was in a different political and economic environment. Increasingly its energies were devoted to responding to national policies over which it had little control and which by the 1990s were posing considerable challenges to the viability of its institutions, most notably its hospitals and postdoctoral programs.

COSTS AND CONTROLS

The passage of Medicare legislation in 1965 led to a significantly greater role of the federal government in financing health care. Millions of Americans over the age of sixty-five now received hospital benefits financed by the federal government. Those enrolled could also participate in a voluntary insurance program to cover physician visits. The passage of Medicaid legislation the same year established a federal-state partnership to provide both hospital and health care provider services to those in poverty, although eligibility, the size of payments to providers, and the range of benefits varied state by state.[1]

One immediate effect of these programs was increased utilization of providers and services by the covered populations. From the first years of its implementation the Medicare program needed vastly greater funding than initially projected. The original legislation mandated that payment to providers be cost-based and retrospective. This meant that the federal government, through its intermediaries and carriers, paid the fees and charges billed by hospitals and physicians with little regard to the appropriateness or the value of the services provided. The Medicare reimbursement formula allowed hospitals to incorporate into their patients' bills capital costs for modernizing or expanding hospital facilities, and this stimulated hospitals to build and grow irrespective of the plans of other institutions or of the needs of the community. Lastly, Medicare would reimburse hospitals for the direct and indirect costs of internship and residency training.[2]

For decades private insurers had used cost-based, retrospective reimbursement methods. Insurers like Blue Cross routinely paid physician fees as long as they fell below the ceiling of what was usual and customary for their specialty and their geographic area. The system discouraged competition among providers. As physicians increased their fees, insurers responded by raising their premiums.[3] The number and percentage of Americans covered by this type of private health insurance increased rapidly beginning in the 1950s. Either employees collectively bargained for these benefits or employers offered them as incentives to retain or attract labor. As those insured no longer paid a significant portion of their health care bills, they had little incentive to seek out lower-cost providers, limit their use of services, or challenge high fees. However, as companies' health insurance costs spiraled, they found an increasing percentage of their revenues being devoted to paying for health care. Business as well as government began looking for ways to restrain this growth.[4]

In the 1970s, most efforts to control these costs took the form of greater regulation and oversight. From 1971 to 1974, the federal government put wage and price controls into place to limit annual increases in hospital expenses and physician fees. This proved to be only a stopgap.[5] In 1972, Congress mandated that Professional Standards Review Organizations (PSROs) be established on the local and state levels throughout the country. These agencies, which had their origin in state government initiatives, were charged with promoting "the effective, efficient, and economical delivery of health care services of proper quality." Focusing on the beneficiaries of the Medicare and Medicaid programs, these agencies, made up of physicians and others, sought ways to reduce the average length of patients' hospitalizations.[6]

In 1974, Congress passed the National Health Planning and Resource Development Act, which created local and state health systems agencies (HSAs) to prevent unnecessary expenditures on hospital facilities. HSAs required any hospital to justify its desired expansion based upon not only its special needs but upon what services neighboring and competing institutions rendered. If successful in the external review process, a hospital would be issued a "Certificate of Need" and the institution could then go forward with its plans. Congress initiated these actions in a legislative climate in which sentiment for national health insurance was growing, and some proponents believed that these measures constituted the initial steps in accomplishing that goal.[7]

PARITY AND INCLUSION

The American Osteopathic Association historically had held significantly different positions on proposed federal health legislation than those embraced by the American Medical Association. For much of the twentieth century, the AMA tried to convince legislators, whatever the content of the proposals and whether or not the association supported them, that DOs were unqualified physicians and that osteopathic schools and hospitals should therefore be excluded from legislative consideration and participation in any programs. While the AMA was traditionally opposed to the federal government's financing or regulating health care services, the AOA was more accepting of federal involvement and took much more moderate positions than did the AMA. To organized osteopathy, the principal concern, whatever the legislation, was not the relative desirability of federal involvement but that DOs have parity with MDs.[8] As legislators proposed new regulatory agencies, the DOs responded as they had traditionally. In their congressional testimony they expressed their concerns over certain provisions in pending bills, pledged to help the programs be successful, and worked to ensure that in whatever bills were enacted DOs would be treated exactly the same as MDs. However, the legislation of the 1970s raised considerable challenges.

With respect to PSROs, the AOA urged its members to monitor developments "to assure the autonomy of the osteopathic profession."[9] As rules were being crafted to implement this program, AOA officials met with representatives of the U.S. Department of Health, Education, and Welfare to ensure DO participation in local and state agencies and to obtain guarantees that, in states with significant numbers of osteopathic physicians, DOs would be responsible for reviewing the practice patterns of other DOs. The AOA also insisted on and won a federal contract to develop model osteopathic hospital admission criteria for use by local PSROs.[10] After these local and state agencies became operational, the AOA successfully lobbied the Bureau of Quality Assurance to allow DOs to secure seats on PSRO governing bodies wherever proper representation was lacking. At the AOA's behest, DOs were also appointed to the National Professional Standards Review Council.[11]

The federal authorization of Health Systems Agencies in 1974 posed similar challenges to the osteopathic profession. Even prior to the act, the AOA had been concerned about discriminatory treatment of DO institutions. Twenty-two states had already enacted "certificate of need legislation," but only two states' laws assured that expansion of osteopathic hospitals would be based only upon the need for osteopathic services and facilities in a community. In other states and under the new federal act no distinctions were made between MD and DO institutions. Thus, protection was given neither to patients who preferred osteopathic care nor to osteopathic undergraduate and graduate student training needs.[12] However, the profession successfully lobbied for changes in a 1979 law revising the National Health Planning and Resources Development Act. As a consequence, wording in the final rules required that the need for services and facilities for osteopathic patients and physicians would be "considered." Though the regulatory language was weak, it did provide osteopathic institutions with a means of appealing denials of desired capital improvements.[13] In addition, state osteopathic associations continued to press their respective legislatures for similar or stronger language in their statutes governing these planning agencies. Successful efforts to establish new hospitals in Tennessee and Florida and approval of the building plans of existing osteopathic hospitals despite opposition of competing MD institutions and local planning agencies would be heralded by the profession.[14]

By the early 1980s legislators recognized that regulatory efforts were not controlling costs. In 1970 the total national expenditures for health care had been $73 billion. By 1980 this figure had more than tripled to $247 billion. During that same period, Medicare and Medicaid costs alone nearly quintupled, from $10.6 to $52 billion. Health care expenditures overall represented almost 10 percent of the gross national product.[15] The

Reagan administration, in response, implemented new initiatives to restrain the rate of growth in Medicare and Medicaid federal spending. Initially, these efforts focused on hospitals. Professional Review Organizations (PROs), which replaced the PSROs, were intended to provide greater oversight of the cost and quality of medical care paid for with federal dollars.[16] In 1983, Congress authorized a new system of Medicare payment to hospitals, based upon diagnostic related groups (DRGs). At the time of hospital admission, each patient would be assessed and categorized by these DRG codes. Hospitals would be paid a flat fee based on diagnosis. If the hospital's cost of caring for a patient was less than the standard reimbursement, the institution pocketed the savings. Conversely, if the cost of care was greater than the standard reimbursement, the institution absorbed the costs. Thus, responsibility for efficient use of services shifted to the hospitals.[17] In addition, Medicare and Medicaid administrators established policies that denied reimbursement for inpatient care that could just as safely and effectively be provided on an outpatient basis. Because fewer patients were being admitted to hospitals and with those admitted being discharged sooner, the competition between institutions to fill beds became more intense.[18]

The private sector was also moving away from the retrospective fee-for-service payment system. In 1973, Congress passed the Health Maintenance Organization (HMO) Act. Based upon models such as the Kaiser-Permanente Plan in California, the HMO delivered health care services to members for a prepaid premium. The amount of the premium for participants would potentially be less than the overall costs of indemnity plans yet cover a similar range of services. Like hospitals under the DRG system, HMOs would bear the financial risks for the care they delivered. To reduce the frequency of serious disease, HMOs emphasized preventive services and encouraged healthy lifestyle behaviors. The patient's primary care physician was to be the HMO's gatekeeper, ideally reducing unnecessary testing and referring patients to costly specialists only when appropriate.[19]

The act as originally crafted by Congress lacked adequate incentives to encourage the development of HMOs. Not until the 1980s and 1990s, as cost reduction pressures mounted, did they grow rapidly. Other alternative organizational forms of what was to be called "managed care" soon followed, including independent practice associations (IPAs), preferred provider organizations (PPOs), and physician-hospital organizations (PHOs). As competition between physicians and practice groups increased, private insurers gained greater leverage and power to limit reimbursement to providers.[20] Because DOs, more than MDs, practiced in nonurban, lower-population areas, managed care did not initially affect many osteopathic physicians. However, by 2000, 74 percent of DOs responding to a national survey reported having managed care contracts.[21]

As these changes took place and particularly as new review and policy setting agencies were created to oversee physician reimbursement, the AOA worked to insure participation in programs and representation in decision making. In the early eighties, the AOA negotiated with the Health Care Financing Administration (HCFA), which administered the Medicare program, to include osteopathic manipulative treatment in the national coding system. In 1988, the association lobbied to require the Harvard researchers working under contract with HCFA to include distinctly osteopathic services in the relative value scale (RVS) they were developing. This project, which assigned values to a wide variety of physician services, would establish the basis for how much physicians would be paid for treating Medicare patients.[22] Inclusion in decision making, however, was sometimes a hard fought struggle. The Physician Payment Review Commission (PPRC), established in 1985 by Congress to advise it on Medicare reimbursement of practitioners, had MD representation but no osteopathic physician members for several years. After an intense campaign of letter writing and lobbying, the first DO representative was appointed to the commission in 1995.[23]

Given the increasing stream of health-related federal legislation, the capacity of the small AOA office in Washington, D.C., to scrutinize every pending bill became a challenge. In 1990, the profession was shocked when it learned that a provision in the just-passed Omnibus Budget Reconciliation Act mandated that, by 1995, only physicians certified by MD specialty boards would be eligible to treat Medicaid-supported pregnant women and children younger than twenty-one. Neither the bill's congressional sponsors nor their staffs intended to exclude DOs; they simply did not know that the osteopathic profession maintained its own specialty boards. Once the bill became law, however, Congress did not pass remedial legislation to correct the error until 1996. Fortunately for DO specialists, the Department of Health and Human Services had agreed to delay the implementation of this provision of the act. As a consequence of that particular lapse in AOA oversight, the association significantly expanded its Washington staff and increased its political presence.[24]

While Congress and federal agencies were generally supportive of equal treatment of DOs and MDs, the AOA found evidence that managed care plans and private insurers discriminated against osteopathic physicians. Some managed care entities refused to hire or contract with DOs unless they had MD specialty board certification. However, in many cases, the AOA found they were able to change existing prohibitions or restrictions simply by providing information to hospitals and managed care organizations on the comparable standards and quality of osteopathic postgraduate training. In other cases, the AOA supported lawsuits brought by individual or groups of DOs to challenge policies making invidious distinctions. State osteopathic associations also lobbied in their respective legislatures for the passage of laws that would prevent hospitals or insurers from discriminating against DOs on the basis of degree and type of specialty board certification.[25]

Many DOs who employed osteopathic manipulative treatment in their practice found it difficult to convince private insurers to adequately compensate them for their services. Some insurers tried to place arbitrary limits on the number of treatments, to lump OMT with nonphysician interventions rendered by chiropractors and physical therapists, or not to pay them anything above a standard office visit that did not include OMT. As these reimbursement problems varied from plan to plan and among assorted jurisdictions and practitioners, the AOA and state osteopathic societies faced a continuing challenge in addressing these problems.[26]

THE ENDANGERED OSTEOPATHIC HOSPITAL

The passage of Medicare directly benefited osteopathic hospitals, and many flourished into the 1970s and early 1980s. In 1975, data collected by the American Osteopathic Hospital Association (AOHA) revealed that osteopathic hospitals provided more than six million days of inpatient care and almost three million outpatient visits annually. Although the average length of stay was shorter by the mid 1970s, outpatient visits were increasing and many hospitals were operating in the black. During the first six months of 1980, revenues at osteopathic hospitals increased by an average of 13 percent over those from the same period the previous year.[27] Despite these numbers, leaders within the profession were becoming anxious about the long-term prospects of their hospitals because of diminishing support by osteopathic physicians and fundamental changes in the methods of payment for hospital care.

Before the 1950s, most osteopathic hospitals had been established to serve the practice needs of a small number of community practitioners. In the mid 1960s, as state medical associations dropped the "cultist" label from DOs and removed any restrictions on their members who associated with osteopathic physicians, DOs were increasingly able to treat patients at traditional allopathic institutions. In the 1970s and 1980s as allopathic hospitals faced a need to fill beds, the DOs, particularly family practitioners, found themselves actively courted by such institutions. These recruitment efforts dismayed osteopathic hospital administrators, who counted on these physicians to keep their own patient census figures high. Compounding the problem faced by osteopathic hospitals, entrenched DO specialists, fearing competition, denied privileges at their particular institutions to other qualified and comparably trained DOs. These shut-out osteopathic specialists reacted by joining nearby allopathic institutions, bringing their patient base with them. As a consequence, osteopathic hospitals suffered.[28]

Both the AOA and the American Osteopathic Hospital Association established working groups to study the problem of decreasing utilization. In 1983, after analyzing questionnaires completed by more than nine hundred osteopathic generalists and specialists and approximately sixty hospital administrators, the Special Committee to Study the Utilization of Osteopathic Hospitals by DOs issued a report. The committee declared that DO general practitioners needed to be accorded the respect and support they deserved from their hospitals and their DO specialist colleagues, that the various sectors of the osteopathic community must communicate more effectively with each other and the public, that DOs must be more loyal to their profession, and that hospitals must upgrade their facilities as well as their support services for DOs.[29]

Some osteopathic hospitals had been able to obtain funding to expand their facilities. In 1978, fifty-six institutions were initiating construction programs that would both add and replace beds. From 1973 through 1983, the number of AOA-accredited osteopathic hospitals with two hundred or more beds increased from twenty-seven to forty-five. But even with expansion, few osteopathic hospitals could match the range or quality of services that could be provided by their much larger, neighboring, and primarily MD-staffed competitors.[30]

With the introduction of DRGs, the patient census of osteopathic hospitals began to decline steadily. In the late 1980s and 1990s cuts in government reimbursement forced hospitals to adopt greater cost-cutting methods, including laying off nurses and ancillary personnel. In addition,

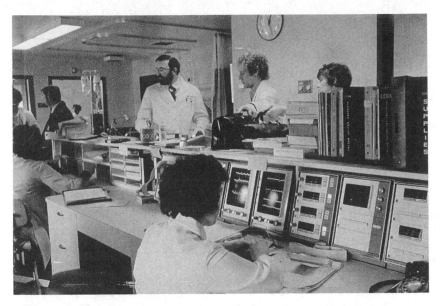

View from nurses' station, Chicago Osteopathic Hospital (1975). *Courtesy of Chicago Osteopathic Medical Institutions (now Midwestern University).*

administrators of several osteopathic hospitals found it difficult to secure managed care contracts and believed that their institutions were the victims of either ignorance or discrimination on the part of the directors of these plans. Osteopathic hospitals were becoming financially distressed.[31]

The location of many of the profession's hospitals was problematic. Osteopathic facilities in cities, such as Detroit, Philadelphia, and Chicago, experienced the effects of demographic change. With middle-class residents moving to the suburbs, the poor became the principal group served by these institutions. Historically, urban osteopathic hospitals relied on the higher rates of reimbursement they received for privately insured patients to compensate for lower-paying government programs covering the poor. As the middle class left the city, however, hospitals could no longer count on private insurers to balance the cost of providing for Medicaid-funded or indigent patients. Rural osteopathic hospitals were also disadvantaged. Under the DRG system they received significantly less reimbursement for services rendered than did their urban counterparts. Small rural osteopathic facilities were unable to secure investment capital to make improvements to attract or retain both physicians and patients. Only osteopathic hospitals located in affluent suburbs weathered the changes generally well.[32]

Not surprisingly, a significant number of osteopathic hospitals closed their doors or ceased to be general inpatient facilities. While it is difficult to trace the fate of all osteopathic hospitals, one important indicator of change is the experience of those institutions accredited by the American Osteopathic Association. The AOA accredited 127 hospitals in 1974. Of these, 96 remained accredited in 1989, but only 59 were still on the rolls in 1999.[33] Some of the once accredited hospitals closed. Others became outpatient facilities, residential treatment centers, or satellite hospitals for larger allopathic institutions.[34]

Some positive developments did emerge from these losses, however. The boards of some of the nonprofit osteopathic hospitals that sold their facilities to multihospital chains used all or most of the proceeds of these sales to create private foundations that would contribute to osteopathic education and other related activities. In 2002, the largest of these, the Osteopathic Heritage Foundation of Columbus, funded principally out of the sale of three Ohio hospitals, had assets exceeding $200 million.[35] In addition, the staff of many osteopathic hospitals that consolidated with or were acquired by larger allopathic hospitals insisted on retaining their osteopathic identity. Often, the acquiring institution agreed to become accredited by the AOA in addition to the allopathic Joint Commission on the Accreditation of Healthcare Organizations. As a consequence, in 1999 the Healthcare Facilities Accreditation Program of the American Osteopathic Association accredited 73 mixed staff hospitals in addition to traditionally osteopathic hospitals, making for a total of 132 institutions in its program, five more than in 1974.[36] Over that twenty-five-year period, the total number of recorded beds and the average size of the hospitals accredited by the AOA actually increased. In 1974, the total number of recorded beds was 18,725, an average of 144 beds per institution; in 1999, the number of recorded beds was 26,875, an average of 203 beds per hospital.[37] Though AOA-accredited mixed-staff hospitals were not specifically osteopathic, they were as a group larger, well-equipped, and offered possibilities for developing new practice and teaching opportunities for the osteopathic profession.

INTERNSHIP AND RESIDENCY SHORTAGES

In the 1950s, there was a widening difference between the allopathic and osteopathic medical professions over the substance of their graduates' first year of postdoctoral education. The number of MD rotating internships (that is, programs in which interns divide their time between several major departments) steadily declined in favor of what were called "mixed" or "straight" internships, which placed emphasis on a specialty. For all practical purposes these latter programs constituted the first year of residency training. In the 1964–65 academic year, 50 percent of all MD postgraduate-year-one (PGY-1) positions were rotating internships; by 1973–74, the percentage had dropped to 19.[38] The mixed and particularly the straight internship allowed the graduate who wanted to specialize to spend more time and thus gain greater experience and skills in his or her chosen field. All AOA-approved PGY-1 programs, on the other hand, whether they were in osteopathic or federal hospitals, continued to be rotating internships. Osteopathic interns had to spend three months in both internal medicine and general surgery and one month in both obstetrics/gynecology and general practice. Students also gained experience in anesthesiology, pathology, pediatrics, and radiology. Continuing AOA support of the rotating internship was rooted in the belief that whether the DO became a primary care practitioner or a specialist, he or she must be prepared to take care of the "entire patient."[39]

In the 1970s, with the establishment of so many new osteopathic colleges, some DOs questioned the capacity of osteopathic hospitals to provide sufficient numbers of internships to meet the expected demand. In 1975, the AOA House of Delegates adopted guidelines submitted by the association's Bureau of Professional Education which would allow "consortia arrangements" wherein two or more smaller osteopathic hospitals could pool their resources to become eligible for interns. Two years later, the House adopted an amended format for the elective rotation of interns through departments of hospitals not accredited by the AOA and for the first time approved the utilization of joint-staff or combined-staff hospitals that were willing to apply for AOA accreditation. These institutions were required to have an adequate DO population in four major departments and to meet other specific criteria. As a result of these changes as well as the establishment of a new formula apportioning internships more on the basis of outpatient services than bed capacity, the AOA set in motion a significant expansion in the number of available internships.[40]

Nevertheless, the marked decline in the number of traditional osteopathic teaching hospitals in the 1980s and 1990s directly affected the profession's ability to keep pace in providing PGY-1 positions for all of its graduates. The AOA's continued commitment to developing new schools and its encouragement of existing private colleges to expand the size of their student bodies greatly exacerbated the problem. Since the AOA did not wish to restrict growth on the undergraduate level, it labored even more intently to increase osteopathic postgraduate educational opportunities.

In 1981, the AOA Task Force on Graduate Osteopathic Medical Education issued a report based on a two-year study funded by the W. K. Kellogg Foundation. Having projected the anticipated rise in DO graduates through 1989, the task force made five recommendations to increase the number of postdoctoral programs under osteopathic auspices: (1) osteopathic hospitals not currently offering internship or residency programs should be given assistance to initiate programs; (2) more mixed-staff hospitals should be encouraged to develop such programs; (3) the number of internships and residencies in existing institutions must be expanded; (4) the criteria by which the number of internship or residency slots was determined needed to be changed; and (5) the AOA should establish a feasibility study to evaluate the efficiency of creating graduate medical education regional consortia. All of these approaches were pursued.[41]

In the mid 1980s the AOA changed its accreditation standards to foster the development of osteopathic college–based programs that could sponsor hospital internships already approved by the allopathic Accreditation Council for Graduate Medical Education (ACGME). The ACGME, which operated under several parent medical groups, including the AMA, had replaced the AMA as the only organization involved in the accreditation of allopathic residencies.[42] Some hospitals staffed primarily by MDs had ACGME-approved PGY-1 positions that had been funded but not filled. Osteopathic colleges approached these hospitals, promoting their graduates as candidates. Many hospitals were pleased that osteopathic colleges could fill these available slots, as they preferred American DOs to international medical graduates (IMGs). Allopathic hospitals met AOA curricular requirements by making small adjustments in the first year of existing family medicine residency or transitional internships. With both the AOA and ACGME independently approving these programs they became known as "dual-accredited internships" By 1997, sixty ACGME-accredited institutions were participating in AOA-approved graduate medical education.[43] As a consequence, DOs expanded their internship base. Between the 1984–85 and 1996–97 academic years, the total number of funded AOA-approved internships jumped from slightly over 1,300 to 1,878—an increase of 44 percent. In 1984–85 there were 200 more new DO graduates than funded internships; by 1989–90, there were more funded internships than graduates.[44]

Although still committed to the rationale behind the rotating internship, the AOA in 1990 finally responded to students who did not want to be family physicians. It approved the development of "specialty track" internships, which served as the first year of residency, and "special emphasis" internships, which did not reduce the length of residency training but provided the postgraduate with more grounding in the respective specialty field. By 2000, 30 percent of DOs in AOA-approved internships were in one or the other of these alternative tracks. While the AOA hoped the dual-accredited as well as the new specialty-oriented internship programs would resolve its PGY-1 shortages, by 1994–95 there once again were more graduates than funded AOA-approved internship slots, and this trend would continue thereafter.[45]

A similar problem existed with respect to osteopathic residency training. By the 1970s a steadily growing percentage of DOs desired residencies, many in primary care areas. The field of "family medicine," which was superseding general practice, required postdoctoral education past the internship. As demand for residency training increased, and with more ACGME programs, not dual-accredited, opening their doors to DOs, leaders within the AOA saw the necessity of having osteopathic and mixed-staff hospitals develop new positions. In the 1980s and 1990s, the AOA approved an impressive number of funded osteopathic residency slots (PGY-2 and above) in solely or dually accredited programs. DOs in AOA-approved osteopathic residencies increased from 699 in 1980–81 to 1,242 in 1985–86, 1,551 in 1990–91, and 2,606 in 1995–96. The specialty fields which attracted the most DO residents were family medicine, emergency medicine, internal medicine, general surgery, obstetrics/gynecology, and orthopedic surgery. Nevertheless, the positions created could not accommodate the needs of all osteopathic graduates. Though the AOA published annual statistics showing many more approved positions than positions filled, a significant percentage of these approved slots were not funded and so existed only on paper.[46]

As a result of these chronic shortages an increasing number and percentage of newly graduating DOs bypassed the AOA intern matching program and elected to enter allopathic first-year residency programs that were not dually approved.[47] MD trainers' experience with the performance of DOs in such programs spread the good reputation of osteopathic graduates as a group. Consequently, ACGME residency program directors, anxious to fill their positions with qualified candidates, dramatically increased the number of mailings to and solicitations of DO students. These efforts were successful. In 1995–96, of the 5,591 DOs in residency training programs, 3,333, or 60 percent, were in either solely or dually accredited ACGME programs. Although a significant number of these DOs had applied for and received AOA approval for such training, a growing number of DOs did not seek the AOA's blessing.[48] Under AOA rules DOs who did not first complete an approved osteopathic internship could not obtain any credit for ACGME residency training toward certification by an osteopathic specialty board. However, DOs completing ACGME residences but lacking an AOA-approved internship were still eligible for and obtained MD specialty board certification. Though these MD-boarded specialists were licensed as DOs, they had little incentive or desire to participate further in the osteopathic community.

As the number of DOs who avoided AOA-approved PGY-1 training grew, the AOA recognized the long-term implications for its membership of maintaining its existing postgraduate education policies. In 1996, the AOA created a mechanism by which DOs in PGY-1 programs solely accredited by the ACGME could petition the AOA for approval of their training. This first year of graduate medical education had to meet essential requirements for the traditional rotating internship and the DO had to document a personal, financial, or legal hardship that explained why the petitioner did not participate in an AOA-accredited program.[49] In 2000, the AOA liberalized its policy further, allowing individual DOs to receive retroactive approval of their ACGME PGY-1 year, thus permitting consideration and approval of their subsequent ACGME residency training. This policy change permitted those DOs the opportunity of securing AOA specialty board certification under rules passed in 1999. In addition, in 2000 the AOA board approved a six-year pilot program that offered approval for the PGY-1 "transitional year" portion of ACGME residency programs that were not dual-accredited but which nevertheless fulfilled the essential requirements of a standard AOA-approved rotating internship. This last change, though it came with restrictions, opened the door wider for senior DO students to sign letters of agreement for solely accredited ACGME training without incurring AOA or osteopathic licensure board penalties for bypassing the AOA intern matching program.[50]

With ACGME programs actively recruiting DO students and the AOA changing its rules to loosen restrictions on their taking solely accredited

ACGME training upon receiving their degree, traditional osteopathic postdoctoral facilities had to develop more-competitive programs if they were to retain a viable number of their graduates. Historically, most osteopathic hospitals operated internships and residencies with little, if any, academic connection to osteopathic medical schools—in contrast to many allopathic programs. ACGME residencies offered their postgraduates a structured, didactic educational curriculum, while osteopathic hospitals were often found wanting in this regard. Unlike ACGME programs, which paid the physician trainers for their services, solely accredited AOA program faculty usually served as unpaid volunteers. Osteopathic educational leaders argued that significant reforms were necessary.[51]

In 1989, the Michigan State University College of Osteopathic Medicine (MSU-COM) established the Consortium for Osteopathic Graduate Medical Education and Training (COGMET). Thirteen Michigan osteopathic hospitals joined with the college through a formal agreement to create a partnership whereby the college would take an active role in enhancing the educational programs and standards within participating hospitals. The goal was to create a seamless and structured curriculum from the first day of medical school through the internship and the residency. The hospitals provided the funding and the college offered in-kind educational services.[52] The Ohio University College of Osteopathic Medicine (OU-COM) soon followed suit, establishing five geographical Centers for Regional Education (CORE), most consisting of multiple hospitals. In the CORE system both OU-COM and the hospitals made significant financial contributions for programmatic development, interactive technology, and the hiring of physician administrators and support staff.[53] Both COGMET and the CORE proved highly successful, not only at retaining their respective students as interns and residents, but in attracting large numbers of DO graduates from other colleges. Their innovative structures and programs drew the attention of educators in other osteopathic schools and in allopathic institutions as well.

In 1995, the AOA, in a bold move, decided that all of its postdoctoral programs should be organized on a consortium basis. It adopted the means to accredit these consortia which it called Osteopathic Postdoctoral Training Institutions (OPTIs). Under the program, any osteopathic or other hospital offering AOA-approved internships and residencies had to become a member of an OPTI. By February 2003, seventeen individual OPTIs had been established, representing all of the fully accredited schools and consisting of all traditional osteopathic teaching hospitals as well as a number of larger allopathic or mixed-staff facilities. Most of these OPTIs have just started their operations and they vary widely in their governance and financing. A small number of OPTIs consist of one or more schools with several osteopathic hospital partners, however, most OPTIs in states or regions without traditional osteopathic hospitals have started from scratch, with the colleges taking the lead in building alliances with allopathic or mixed-staff institutions. Most OPTIs are currently small and can provide only a fraction of the number of internship and residency slots needed by the graduates of the osteopathic college or colleges participating in that OPTI.[54] Consequently, there has been a steady increase in the percentage of osteopathic graduates entering solely accredited ACGME programs.

Furthermore, the challenge of providing a sufficient number of high-quality postdoctoral positions under AOA auspices is becoming more difficult as the federal government changes the basis upon which it will fund graduate medical education. Under the provisions of the Balanced Budget Act of 1997, Medicare-based graduate medical education funding was drastically cut. But the most significant feature of the legislation was that it froze the number of fundable MD and DO internships and residencies in each hospital to those filled in the prior year, as a way of eliminating excess positions. No distinction was made between the internship-and-residency-starved traditional osteopathic program and the postgraduate program surpluses generally found in allopathic medicine.[55] As a result, unless Congress makes significant changes in the law to accommodate the special circumstances and needs of osteopathic medicine, the OPTI program can only flourish if it finds alternative and additional means to finance existing positions in currently affiliated hospitals, attracts community hospitals that under the Balanced Budget Act are still eligible to create new federally funded graduate programs, and affiliates with a far larger number of hospitals with ACGME programs to create substantially more dual-accredited programs.[56]

THE CHALLENGE OF DISTINCTIVENESS

Osteopathic medicine occupies the same professional space as its older, larger, and more socially dominant counterpart, which wishes to absorb it. Given its increasing closeness in standards and services to its dominant rival and the greater association between the practitioners of both professions, it makes little sense for the osteopathic profession, if it wishes to retain its independence, to continue stressing its similarities with allopathic medicine. The public is unlikely to believe that DOs can ever practice allopathic medicine in all of its manifestations as well as can MDs. Nor does professional mimicry appear to be a viable way of obtaining public favor or recognition. As is readily apparent by the billions patients spend on the many forms of alternative medicine, they want choices. Therefore, from a market perspective, osteopathic medicine should find and develop the resources to produce not only qualified physicians but practitioners widely perceived by the public and themselves to be different from MDs and arguably better in some aspects of the way they care for patients. However, given the state of the osteopathic profession's current infrastructure, funding sources, and the degree to which distinctiveness is now practiced and taught, this will be no easy challenge.

OSTEOPATHIC COLLEGES

Osteopathic medicine is now the fastest growing segment of the U.S. physician and surgeon population (see Table 5). In 1962, just after the California merger, there were approximately 11,000 DOs in practice. In 2002, there were 47,000. Based on current trends, by 2020 there will be approximately 80,000 DOs. There are now 15 active DOs per 100,000 Americans compared to 6 per 100,000 forty years ago. In 2000, office-

Table 5. Distribution of Active DOs and MDs by State, 1999

State	Total Number Active Physicians	MDs	DOs	Percentage of DOs among Active Physicians	Percentage Change in DOs since 1990
Alabama	8,720	8,449	271	3.1	156
Alaska	1,210	1,133	77	6.3	57
Arizona	10,588	9,529	1,059	10.0	52
Arkansas	4,974	4,810	164	3.2	148
California	79,403	77,407	1,996	2.5	75
Colorado	10,339	9,672	667	6.5	51
Connecticut	11,377	11,181	196	1.7	145
Delaware	1,875	1,708	167	8.9	38
District of Columbia	3,890	3,859	31	0.8	-18
Florida	38,117	35,469	2,648	6.9	60
Georgia	16,812	16,273	539	3.2	83
Hawaii	3,335	3,214	121	3.6	66
Idaho	2,013	1,914	99	4.9	102
Illinois	31,757	30,079	1,678	5.3	62
Indiana	11,847	11,248	599	5.1	87
Iowa	5,687	4,792	895	15.7	38
Kansas	5,838	5,306	532	9.1	45
Kentucky	8,302	8,086	216	2.6	108
Louisiana	10,503	10,403	100	1.0	85
Maine	3,225	2,806	419	13.0	34
Maryland	19,888	19,534	354	1.8	95
Massachusetts	24,052	23,708	344	1.4	94
Michigan	25,557	20,873	4,684	18.3	28
Minnesota	11,855	11,635	220	1.9	100
Mississippi	4,733	4,528	205	4.3	153
Missouri	13,713	12,083	1,630	11.9	19
Montana	1,752	1,675	77	4.4	93
Nebraska	3,611	3,544	67	1.9	81
Nevada	3,408	3,151	257	7.5	127
New Hampshire	2,844	2,748	96	3.4	191
New Jersey	25,384	22,959	2,425	9.6	48
New Mexico	3,885	3,704	181	4.7	23
New York	67,823	65,453	2,370	3.5	106
North Carolina	17,702	17,395	307	1.7	201

continued

Table 5. Continued

State	Total Number Active Physicians	MDs	DOs	Percentage of DOs among Active Physicians	Percentage Change in DOs since 1990
North Dakota	1,440	1,383	57	4.0	159
Ohio	28,632	25,484	3,148	11.0	32
Oklahoma	6,676	5,499	1,177	17.6	34
Oregon	7,683	7,317	366	4.8	42
Pennsylvania	37,957	33,263	4,694	12.4	34
Rhode Island	3,358	3,188	170	5.1	34
South Carolina	8,251	8,070	181	2.2	135
South Dakota	1,465	1,401	64	4.4	94
Tennessee	13,453	13,141	312	2.3	126
Texas	42,122	39,565	2,557	6.1	40
Utah	4,259	4,149	110	2.6	189
Vermont	1,816	1,772	44	2.4	47
Virginia	16,995	16,550	445	2.6	96
Washington	13,822	13,286	536	3.9	40
West Virginia	4,293	3,836	457	10.6	71
Wisconsin	12,096	11,661	435	3.6	52
Wyoming	852	819	33	3.9	106
Total	701,189	660,712	40,477	4.8	

Source: "AOA Fact Sheet" (August 2000): 29; Thomas Pasko, Bradley Seidman, Scott Birkhead, "Physician Characteristics and Distribution in the U.S., 2001–2002 Edition" (Chicago: AHA Press, 2002).

based DOs received 66.7 million patient visits or 24.3 visits per 100 persons in the United States. DOs are presently involved in providing for the health care needs of as many as 30,000,000 Americans.[1] Despite their growing numbers, DOs are not evenly distributed throughout the United States. In many parts of the country there are so few DOs that the profession is socially invisible. Michigan and Pennsylvania have the most practitioners, with more than 4,500 active DOs apiece; ten other states each have more than 1,000 DOs and DOs in these twelve states comprise approximately 77 percent of active osteopathic physicians and surgeons not in military service. Nevertheless, in the past dozen years the largest percentage gains of DOs have been in those states with smaller osteopathic representation.[2]

This recent rapid growth has been fueled by the development of additional schools and expansion of the class size in longstanding as well as newer institutions. In 1962, there were five osteopathic colleges, in 1982 fifteen. In 2002, 19 of 144 U.S. medical schools were osteopathic institutions. In the past twenty years new private DO-granting schools were established in Florida, Arizona, Kentucky, Pennsylvania, and California, and in 2003 an osteopathic college in Virginia admitted its first class. Between 1962 and 2002 DO graduates have increased sevenfold, and almost 14 percent of all U.S. medical school graduates are now DOs. Plans are being developed to establish additional osteopathic colleges.[3]

All this occurred despite the recommendations of most health workforce experts, who argued that the United States needed fewer not more physicians. The leadership of the AOA pointedly countered that contention, saying that, given the number and percentage of DO graduates who enter primary care fields and practice in underserved areas, osteopathic colleges have done a better job than either U.S. or foreign MD-granting schools of providing the types of physicians this country needs.[4] According to a 1998 study by the American Medical Student Association, of all U.S. medical schools the top twelve producing the highest percentage of graduates entering primary care residencies were all DO-granting schools, with the other four DO institutions in the survey falling among the top twenty. This is not surprising, since osteopathic colleges focus on primary care in student recruitment, curriculum, role models, and opportunities. In 1999, approximately 60 percent of all active DOs were in

Table 6. Comparisons of DO and MD Schools and Graduates

Year	MD Schools	DO Schools	Percentage DO Schools	MD Graduates	DO Graduates	Total Graduates	Percentage DOs
1962	82	5	5.7	7,168	362	7,530	4.8
1972	89	7	7.3	10,396	649	11,045	5.9
1982	121	15	11.0	16,012	1,317	17,329	7.6
1992	125	15	10.7	15,365	1,606	16,971	10.4
2002	125	19	13.2	15,810	2,543	18,353	13.9

Sources: Journal of the American Medical Association 288 (September 2002); The DO 43 (August 2002): 54; Allen M. Singer. "2000 Annual Statistical Report" (Chevy Chase, AACOM, 2001), AAMC Data Book (Washington D.C.: AAMC, 2000).
Note: A new MD college in Florida and a new DO college in Virginia had preaccredidation status in 2002 but did not admit their first class until 2003.

primary care (48 percent in family medicine, 8 percent in internal medicine and 3 percent in pediatrics).[5]

Osteopathic colleges are not research-oriented institutions. Indeed, unlike most MD schools, which boast of substantial numbers of faculty and research grants, DO schools have comparatively few full-time faculty members and their mission is service directed.[6] Many osteopathic faculty members have been hired and promoted on their ability to teach a broad range of curricular subjects rather than on their research credentials. Thus, while contributing a significant number of graduates to areas of the physician workforce that MD schools have insufficiently addressed, osteopathic schools have made comparatively little contribution to creating new knowledge or producing graduates who will be active researchers. Indeed, one recent study concluded that if all nineteen osteopathic schools were treated as one institution, it would rank 202nd in National Institutes of Health funding.[7]

Within osteopathic undergraduate education there are notable funding disparaties between the six public and the thirteen private osteopathic schools. In the 2000 fiscal year, the average current fund revenue for the six public colleges was $48.5 million per school, compared to an average of $32.6 million per private institution. The private colleges are heavily dependent upon tuition, which constitutes an average of 70 percent of their current fund revenues, compared to 12 percent for the public schools. The state schools, on the other hand, draw 39 percent of their revenues from legislative appropriations and 30 percent from practice plans. Unlike MD schools, neither public nor private DO colleges have attracted significant philanthropic support or developed substantial endowments.[8]

The state schools, given their broader sources of funding and their legislative mandates, have limited and kept stable their enrollments, which in 2000–2001 averaged 101 new students. The private colleges, so reliant upon tuition, averaged 178 new students and have used an increase in the number of matriculants as well as increases in tuition as the most dependable ways of generating additional revenue to support their educational programs. The average number of full-time faculty in public schools is more than double that found in the private colleges, making for a more favorable faculty-student ratio. As a consequence of their more limited resources, private colleges are more likely to rely on part-time and voluntary instructors, to depend upon large lecture formats, and to use web-based self-study technology to deliver their curriculum.[9]

In the 1980s and 1990s several of the once freestanding private osteopathic schools expanded their educational mission and evolved into "health science universities," establishing accredited colleges of pharmacy, podiatry, physical therapy, optometry, dentistry, and physician assistant programs. This expansion appears to have increased the overall economic viability of the resulting institution through cost-sharing of faculty, support staff and facilities. Some of these health science universities have been able to raise the capital needed to erect large and modern buildings specifically for osteopathic education, and DO students now get to interact and in some instances take classes with other health professionals in training.[10]

Ironically, the public osteopathic schools, though more adequately funded, have faced a greater challenge to their continued existence than the private colleges. Governors and state legislatures have periodically threatened their respective schools with closure or consolidation, under the pressure of chronic shortages in state revenues and the perception of an oversupply of physicians. However, each of the public osteopathic colleges has a stronger record than their MD college counterparts in their state in producing graduates who remain within its borders, enter primary care, and practice in underserved, particularly rural, areas. As a result, osteopathic colleges have successfully fought such efforts, and in general the resulting review process strengthened their standing and reputation among lawmakers.[11]

The educational credentials of matriculants to osteopathic schools continue to improve. Osteopathic students enter with baccalaureate degrees, some with advanced training, and most have graduated from their undergraduate college or university in the top 25 percent of their class. MCAT scores on average, however, are significantly lower than matriculants at MD schools, although they are likely to be more consistent with the scores of allopathic students who enter primary care careers. Osteopathic students tend to be slightly older than their MD counterparts, because a greater percentage of these matriculants enter school after having pursued another career. A majority of students who enroll in osteopathic colleges are quite knowledgeable about the profession—many having had a DO as a physician. However, a significant minority enter osteopathic schools after having been unsuccessful, despite good academic credentials, in gaining acceptance at an MD school. DO-granting colleges therefore face a special challenge acculturating these "second-choice" students with osteopathic beliefs and practices.[12]

U.S. Colleges of Osteopathic Medicine, 2003

Code	Name of College	City
AZCOM	Arizona College of Osteopathic Medicine of Midwestern University	Glendale, AZ
CCOM	Chicago College of Osteopathic Medicine of Midwestern University	Downers Grove, IL
DMU/ COMS	Des Moines University / College of Osteopathic Medicine and Surgery	Des Moines, IA
KCOM	Kirksville College of Osteopathic Medicine, A. T. Still University	Kirksville, MO Erie, PA
LECOM	Lake Erie College of Osteopathic Medicine	
MSUCOM	Michigan State University College of Osteopathic Medicine	East Lansing, MI
NSUCOM	Nova Southeastern University College of Osteopathic Medicine	Fort Lauderdale, FL
NYCOM	New York College of Osteopathic Medicine of the New York Institute of Technology	Old Westbury, NY
OSU-COM	Oklahoma State University Center for Health Sciences College of Osteopathic Medicine	Tulsa, OK
OU-COM	Ohio University College of Osteopathic Medicine	Athens, OH
PCOM	Philadelphia College School of Osteopathic Medicine	Philadelphia, PA
PCSOM	Pikeville College School of Osteopathic Medicine	Pikeville, KY
TUCOM	Touro University College of Osteopathic Medicine	Vallejo, CA
UHS-COM	University of Health Sciences-College of Osteopathic Medicine	Kansas City, MO
UMDNJ-SOM	University of Medicine and Dentistry of New Jersey-School of Osteopathic Medicine	Stratford, NJ
UNECOM	University of New England College of Osteopathic Medicine	Biddeford, ME
UNTHSC/ TCOM	University of North Texas Health Science Center at Fort Worth / Texas College of Osteopathic Medicine	Fort Worth, TX
VCOM	Edward Via Virginia College of Osteopathic Medicine	Blacksburg, VA
WU/COMP	Western University of Health Sciences / College of Osteopathic Medicine of the Pacific	Pomona, CA
WVSOM	West Virginia School of Osteopathic Medicine	Lewisburg, WV

Interactions between members of DO and MD groups have become significantly less contentious in recent decades. Many state osteopathic associations work with their allopathic counterparts to lobby for legislation such as malpractice tort reform, public health initiatives, and combating the efforts of nonphysician providers to expand their scope of practice. On the national level, the AOA and the American Association of Colleges of Osteopathic Medicine (AACOM) are members, along with the AMA and the American Association of Medical Colleges (AAMC), of various coalitions of health-related interest groups. These osteopathic and allopathic associations communicate regularly about pending federal legislation. Nevertheless, allopathic organizations and prominent MDs continue to question the independence of the osteopathic profession. For example, at two sets of well publicized meetings in the mid 1990s between DO and MD organizational leaders sponsored by the Josiah Macy Jr. Foundation, several MD representatives opined that, given the great similarities now existing between the two professions, they could see no reason why there should not be only one great united house of medicine.[13]

Some MD groups, however, more directly challenge osteopathic autonomy, most notably with respect to licensure. Some state medical associations periodically support legislation to eliminate independent osteopathic licensure boards, particularly in those states with smaller osteopathic representation.[14] On the national level, MD members of the Federation of State Medical Boards launched a campaign in 1998 questioning the existence of a separate osteopathic pathway to licensure. For decades the federation, made up primarily of members of state medical licensing boards, has had as its goal the creation of a uniform single examination to license physicians in all states. In the process it has championed the test of the National Board of Medical Examiners (NBME) known as the United States Medical Licensing Examination (USMLE), which was designed to assess the qualifications of MD candidates. Nevertheless, all states—except Louisiana—also accepted the test of the National Board of Osteopathic Medical Examiners (NBOME) as an equivalent examination for assessing DOs. When the NBOME developed a new, and arguably better, test for DOs called the Comprehensive Osteopathic Medical Licensing Examination-USA (COMLEX-USA), many MD members of the federation took the opportunity to challenge both the principle of a separate osteopathic licensing mechanism and the validity of the new exam. With federation members demanding that the NBOME provide sufficient evidence that the COMLEX-USA was the equal of the USMLE, the NBOME met the standard; and in 2001 its opponents withdrew their objection to COMLEX-USA, though not their long-stated goal of a uniform pathway to licensure. Shortly thereafter, the state of Louisiana changed its requirements, making COMLEX-USA an acceptable test for licensure of DOs in all states.[15]

Despite these interprofessional conflicts, there is now widespread MD acceptance of DOs as colleagues, and this acceptance presents a real dilemma for organized osteopathy. For most of its history the AOA derived a great deal of internal cohesion and social solidarity among its membership from the menacing actions of the once-powerful AMA. Older DOs, who fought the long and sometimes brutal battle to achieve recognition and equality, observe that where the AMA once used every opportunity to "kill osteopathic medicine," now organized medicine just wants to "love us to death." Indeed, the AMA has abandoned its forced amalgamation efforts, partly because they didn't work and partly because they likely violated anti-trust laws.[16] Instead, the AMA has tried to build bridges. It recently designated two seats at its House of Delegates for representatives of the AOA—but to date the AOA has rejected the offer and the seats have gone unfilled.[17]

What the AMA once tried to achieve through organizational amalgamation, it now hopes to accomplish through individual assimilation. DOs and MDs increasingly practice together in the same hospitals and in medical groups, and by and large enjoy cordial relations. The fact that approximately 60 percent of all osteopathic residents are now enrolled in programs approved by the Accreditation Council for Graduate Medical Education is the strongest evidence that this process of assimilation will continue. This trend is especially troubling to the AOA. The prospects for the allegiance of ACGME-trained DOs, particularly those who bypassed AOA approval, are not promising. Indeed, a sizable number of these practitioners have instead joined the AMA and identify more with their allopathic colleagues. As of November 2002, 7,936 DOs (or 17 percent of all active osteopathic physicians) were AMA members. Nevertheless, the AOA continues to draw significant support among DOs. Overall, the percentage of osteopathic physicians who have joined the AOA has remained stable in recent years and in 2002 stood at 63 percent. On the other hand, the AMA, despite an infusion of osteopathic physicians into its ranks, has been struggling to enlist new members and now represents only 30 percent of all active post-residency MDs.[18]

OSTEOPATHIC PRINCIPLES AND PRACTICES

The AOA is strongly committed to maintaining the osteopathic profession's independent status, and there is no evidence that it will change that position in the foreseeable future. But if it is to remain independent and to flourish, osteopathic medicine must both create a clearer vision of its role and develop a recognizable identity. Having achieved legal and professional equality, a growing number of DOs are urging their colleagues to reexamine their roots. Leaders of the AOA increasingly highlight the value added to patient care by traditional osteopathic philosophy. They seek ways to strengthen the teaching of distinctive osteopathic diagnostic and therapeutic practices in osteopathic colleges and hospitals and dual-accredited graduate programs and to promote their utilization in osteopathic practice. Ironically, after decades of striving to convince legislators and the public of the close similarities between DOs and MDs, many within the profession now see the importance of stressing the differences between the two types of practitioners.[19]

Osteopathic medicine is a social movement as well as a profession. As a social movement it espouses a philosophy of medicine and a set of principles that distinguish it from its allopathic counterpart. Indeed, extolling the virtues of a medical philosophy at all makes the osteopathic profession different. Although policymakers, social scientists and others refer to the MD profession as "allopathic medicine," the term itself is an historical artifact not reflecting any body of beliefs embraced and shared by its own members. For well over a century the MD profession has pointedly rejected the adoption of any philosophical belief system governing health and disease, equating philosophy with dogma and arguing that its professional approach to medicine is dependent solely upon scientific evidence.

In the 1920s the faculty at the Kirksville school codified a set of fundamental osteopathic principles that were widely accepted throughout the profession. These were somewhat revised at midcentury. Four tenets were enunciated: first, the body is a unit and the person represents a combination of body, mind, and spirit; second, the body is capable of self-regulation, self-healing, and health maintenance; third, structure and function are reciprocally interrelated; and fourth, rational treatment is based upon an understanding of body unity, self-regulation, and the interrelationship of structure and function.[20] Several authors have explained how these particular principles both link the osteopathic profession to conventional medicine and provide what they believe is a more holistic, patient-centered approach. Embedded in the philosophy is a rationale for the incorporation of osteopathic manipulative treatment. The neurophysiologist Irwin Korr put forward four related propositions in this regard. First, he noted, the vertical human framework is highly vulnerable to gravitational, torsional, and shearing forces. Second, since the massive, energy-demanding musculoskeletal system has rich two-way communication with all other body systems, it is, because of its vulnerability, a common and frequent source of impediments to the functions of other systems. Third, these impediments exaggerate the physiological impact of other detrimental factors in the person's life, and, through the central nervous system, focus that impact on specific organs and tissues. Fourth, the musculoskeletal impediments or somatic dysfunctions are readily accessible to the hands and responsive to manipulative treatment and other methods developed and refined by the osteopathic medical profession.[21]

Whatever the rationale, the use of osteopathic manipulative treatment in overall patient management has declined. In 1972 the independent journal *Osteopathic Physician* published the results of a mailed questionnaire returned by 234 DOs located in ten states. Asked the question "On what percentage of your patients do you make use of manipulation," 66 percent replied, "Less than 50 percent" and 37 percent responded, "Less than 20 percent."[22] In the 1974 national ambulatory medical care survey carried out by the National Center for Health Statistics, it was estimated that fewer than 17 percent of all patient visits to office-based DOs included osteopathic manipulative treatment.[23] More recent studies by Fry (1996), Johnson, Kurtz, and Kurtz (1997), and Aguwa and Liechy (1999) all confirm a continuing downward trend.[24]

In the latest study by Johnson and Kurtz (2001), 30 percent of 375 osteopathic family physicians surveyed reported that they employed OMT on "less than 5 percent" of their patients, 50 percent reported using OMT on "from 5 to 25 percent" of patients, and only 20 percent used it on "more than half" their patients. Only 30 percent of 580 surveyed specialists reported using OMT on "more than 5 percent" of patients. Interestingly, the great majority of all DO respondents had favorable attitudes towards OMT; 96 percent agreed or strongly agreed that it is an efficacious treatment. Over 70 percent agreed or strongly agreed that they personally received OMT or provided it to friends, colleagues, or relatives outside their practice.[25] The factors explaining the lessened use of OMT on patients include the diminished number of hours spent on osteopathic diagnosis and treatment in the undergraduate curriculum, the greater emphasis given to other modalities, restriction of opportunities for use in clerkship and postgraduate settings, poor or no reimbursement for distinctly osteopathic procedures, and increasing percentages of new graduates going into specialties in which the use of OMT is not regarded as necessary.

During the first two years of osteopathic school, students receive an average of 218 contact hours in osteopathic principles and practices (OP&P). This time constitutes approximately 24 percent of the clinical science curriculum and 12 percent of the total curricular hours for those years.[26] Students spend some of these OP&P hours in lecture halls, but they are primarily in clinical labs, that is, large rooms with treatment tables for teams of students led by one faculty member with several other faculty and student fellows serving as table-trainers. In this setting the students learn both structural diagnostic methods and manipulative techniques. Students take turns being the "operator" and the "patient." Of all subjects students take, this is the most critical in developing an osteopathic identity. The degree to which students grasp the basic principles, see the utility of the methods taught, and gain palpatory literacy may largely shape their future professional choices. This is especially true for the significant minority for whom an osteopathic school was a second choice. The Educational Council on Osteopathic Principles of the AACOM has worked to standardize the curriculum, osteopathic techniques utilized, and descriptive nomenclature. The publication, under the auspices of the AOA, of a comprehensive introductory textbook, entitled *Foundations for Osteopathic Medicine*, has also helped to standardize what is taught. The text, now in a second edition, has been adopted by every osteopathic school as either required or recommended student reading.[27]

The colleges vary appreciably, however, in terms of the fiscal and faculty resources devoted to the teaching of osteopathic manipulative medicine (OMM). Some schools, such as the private University of New England College of Osteopathic Medicine, built beautiful and well-equipped OMM labs that are the centerpiece of the institutions. In addition, some colleges employ several full-time OMM faculty members. Other schools, facing chronically tight budgets, are more dependent upon part-time and volunteer practitioners to deliver the OMM curriculum, and have labs that are in serious need of updating.

All students in the first two years of their education obtain a basic grounding in osteopathic fundamentals. However, there is often inadequate attention to further developing what skills students have already mastered in their third- and fourth-year clinical rotations. Although in many osteopathic hospitals students will routinely perform an osteopathic structural examination as part of the overall history and physical, they generally do not record findings of somatic dysfunction on the chart, and they administer manipulative treatment only to a small number of hospitalized patients. Some trainers and students regard palpation-based structural findings as not directly germane to their patients' problems and believe that manipulative procedures are more appropriate to a primary care office setting. Whatever the relative merits of their arguments, if administrators and preceptors at the clinical sites are not committed to promoting the diagnostic or therapeutic distinctiveness of the profession, then whatever skills the student has previously learned simply wither.[28]

The newly developed Osteopathic Postdoctoral Training Institutions provide osteopathic colleges with the means to introduce and incorporate a formal didactic curriculum in OP&P during the third and fourth years and into graduate education. Some college departments of osteopathic principles and practices have developed innovative programs in their OPTI that their proponents regard as successfully addressing this problem. However, a significant number of third- and fourth-year clerkship sites are at hospitals that are not part of OPTIs, and several OPTIs have had great difficulty adhering to AOA accreditation rules or abiding by their own organizational bylaws with respect to the integration of osteopathic principles and practices during the internship and residency years.[29]

Public demand for the services of "ten-fingered" DOs is reportedly high in many localities as is patient satisfaction with osteopathic manipulation.[30] Whether younger DOs employ palpatory diagnosis and manipulative treatment in their practice depends in part upon how well they are compensated for integrating these methods. Some private insurers and Medicare will reimburse for both osteopathic evaluation of somatic dysfunction and the use of manipulative intervention. Nevertheless, the fact that these procedures may add time and cost to patient visits can make third party payment problematic, and DOs have often had to justify to others the inclusion of these procedures that, for them, constitute fundamental cornerstones in the practice of their profession.

Third party payers want evidence that a distinctly osteopathic approach works for given complaints and that it is cost effective. Consequently, the osteopathic profession will be under increasing pressure to demonstrate that its distinctive health care services add value to patient encounters. But third parties are not the only ones calling for documentation. DO students and graduates over the decades have consistently urged the profession to underwrite and conduct research studies carefully and objectively examining the clinical significance of "somatic dysfunction" and the relative value of osteopathic manipulative treatment.

OSTEOPATHIC CLINICAL RESEARCH

Although several osteopathic schools have produced empirical studies related to distinctive osteopathic practices since the 1960s, the most significant and sustained efforts until the late 1990s occurred at the Michigan State University College of Osteopathic Medicine, performed by William Johnston, DO (1921–2003) and his colleagues, and at the Chicago College of Osteopathic Medicine, now a component of Midwestern University. Scientists at the Kirksville College had examined the neurophysiologic basis of the "osteopathic lesion," what became known as "somatic dysfunction"; now investigators at these two schools researched the clinical aspects of the phenomenon.[31]

Currently, somatic dysfunction is defined as "impaired or altered function of related components of the somatic system: skeletal, arthroidal, and myofascial structures, and related vascular, lymphatic, and neural elements." The definition further specifies that the positional and motion aspects of somatic dysfunction are best described using at least one of three parameters: the position of the body part as determined by palpation and referenced to its adjacent defined structure, the directions in which motion is freer, and the directions in which motion is restricted.[32]

Over the past twenty years there have been three broad lines of inquiry regarding somatic dysfunction: the establishment of interexaminer agreement of findings, the measurement of somatic dysfunction through instrumentation, and the documentation of clinical correlations between somatic dysfunction along the spinal column and conditions elsewhere.[33] This research has yielded evidence that, if appropriate training is provided, several examiners can achieve through palpation a high degree of agreement on the presence or absence of somatic dysfunction at specific spinal locations.[34] Researchers have also shown that electromyography can provide instrument-based evidence confirming palpatory findings of somatic dysfunction.[35] In addition, several investigators have associated patterns of somatic dysfunction with cardiovascular, pulmonary, renal, and mental diseases.[36] Remarkably, though, in the past two decades, no articles have been published in the *JAOA* that empirically test whether somatic dysfunction as specifically and objectively identified along the spinal column can be eliminated through the use of osteopathic manipulation and whether such treatments are correlated in any way with demonstrable physiological changes elsewhere in the body. Such studies are absolutely essential to testing the fundamental premises upon which the profession rests.

In the early 1990s, the AOA Bureau of Research decided that clinical investigations under osteopathic auspices which showed good results for osteopathic manipulative medicine would either be open to question or simply ignored by a nonosteopathic audience. Consequently, using several hundred thousand dollars raised through a special assessment on the general membership, the AOA funded a prominent and experienced MD investigator at Rush University Medical School in Chicago to develop and conduct a study based in Rush's facilities directly comparing the outcomes of MDs and DOs in treating lower back pain. The results were published in an article in the *New England Journal of Medicine* in 1998.[37]

In this study researchers randomly assigned to treatment groups 178 subjects who had had back pain for at least three weeks but not as long as six months. All patients were treated either with one or more standard medical therapies or with osteopathic manipulative treatment. The results showed that there were no statistically significant differences between the two groups in any of the primary outcome measures. However, the osteopathic treatment group required significantly less medication (analgesics, anti-inflammatory agents, and muscle relaxants) and used less physical therapy.

These published results were interpreted along predictable lines. MDs and some DOs argued that since the findings showed no significant difference in primary outcome measures, distinctly osteopathic medical intervention had not made any improvement over conventional medicine in patient care. However, the AOA and many DOs claimed that this study had in fact demonstrated that patients could be safely and effectively treated through osteopathic manipulation with less exposure to pharmaceuticals and their associated side effects. As much attention was drawn by both sides to the underlying political implications of the study—that is, to the question of whether osteopathic medicine should remain an independent profession—as to the perceived or actual methodological shortcomings of the research.[38]

Needing to facilitate more varied studies on osteopathic manipulative treatment, the profession has once again turned to funding its own investigators and trying to build a credible research base in osteopathic colleges. The AOA has annually provided almost one-half million dollars in a competitive application process to support pilot studies, some of which have eventuated in reports published in the *JAOA*. In recent years osteopathic researchers, with or without AOA funding, have, for example, compared thoracic manipulation with incentive spirometry in preventing postoperative atelectasis, examined the effect of suboccipital dermatomyotonic stimulation on digital blood flow, measured the effect of lymphatic and splenic pump techniques on the antibody response to hepatitis B vaccine, examined whether osteopathic manipulative treatment improves gait performance in patients with Parkinson's Disease, explored whether OMT benefits elderly patients with pneumonia, and considered whether OMT in addition to medication relieves pain associated with fibromyalgia syndrome.[39] Unfortunately, to date pilot projects under osteopathic auspices have rarely led to larger and more scientifically rigorous studies.

In 2001, the profession embarked on a more ambitious program. The AOA, AACOM, and the American Osteopathic Foundation pledged to contribute $1.1 million to one school—the Texas College of Osteopathic Medicine at University of North Texas—which has significantly increased its external research funding in recent years. Designated by the AOA as the Center for Osteopathic Research and Excellence these osteopathic granting agencies hope that their modest seed money will lead to federal support for this center as research projects multiply and come to fruition.[40] Increased public or private funding at this school or elsewhere will likely be dependent upon the degree to which osteopathic researchers can demonstrate that they can conduct and publish controlled outcome studies with large sample sizes that can definitively answer research questions. The ability of the profession to show conclusively that osteopathic management has benefit to patients with specific conditions would add significantly not only to the scientific reputation of osteopathic medicine but also, and just as importantly, to its public visibility.

VISIBILITY AND RECOGNITION

Osteopathic medicine may well be the best-kept secret in American health care. After more than a century of existence, the osteopathic medical profession is the least known of the major health care professions in the United States. Separate studies commissioned in the late 1990s by the AOA and AACOM reveal that less than 15 percent of Americans know the scope of osteopathic medical licensure and can articulate meaningful differences between DOs and other health care practitioners.[41] This problem can be attributed to their comparatively small numbers, their strength in the relatively unglamorous field of primary care, and the fact that their schools and hospitals are not research-oriented institutions. DOs are also underrepresented in the overall physician population in the national media capitals of New York, Los Angeles, and the District of Columbia.

The sheer numbers of allopathic physicians, dentists, and nurses reflect the widespread visibility and social dominance of these health care professionals. They have carved out fields and activities that are generally recognized as falling appropriately within their domains. But large numbers do not tell the whole story. The public recognizes the much smaller health care professions of optometry, podiatry, and chiropractic, and most people have a basic understanding of the range of services that the members of these smaller professions provide. Even more striking is that extremely small groups of health care providers, notably acupuncturists and homeopathic physicians, because of their marked differences in beliefs and practices from those of conventional medical providers, are better known for what they do than are osteopathic physicians.

One of the great ironies of osteopathic medicine's development is that, as the profession broadened its curriculum and obtained for its practitioners the same legal practice privileges as allopathic physicians, DOs became less distinguishable from MDs. Thus, even though the number of osteopathic physicians has dramatically increased, the profession remains socially invisible. This social invisibility has had numerous and severe consequences for the profession. On the societal level, osteopathic invisibility has led federal and state lawmakers to craft health care bills that either ignore osteopathic medicine or do not take into consideration its special circumstances and needs, health care planners to overlook osteopathic physicians in their forecasts, and health reporters to either be unaware of or ignore the osteopathic profession in stories to which the profession's positions or role in health care are relevant.[42] On the individual level, DOs experience their profession's social invisibility first hand. A practitioner who identifies him or herself to strangers as a DO or as an osteopathic physician usually receives the response "What's that?" Even after being told what a DO is, strangers are likely to ask additional questions,

Bringing health care to rural Ohioans (2000).
Courtesy of Ohio University College of Osteopathic Medicine.

such as "How does a DO differ from an MD?" and "How does a DO differ from a chiropractor?" From the time they begin their medical education, osteopathic students know that they have to contend with this identity problem.[43]

During the past several decades, the profession has tried various means of educating the public. These have included such standard public relations methods as distributing news releases; forming speakers' bureaus; encouraging newspapers, magazines, and television programs to do news and human interest stories on DOs; producing short films, audiotapes, and public service announcements; publishing pamphlets for osteopathic physicians' offices; and placing advertisements in journals and magazines targeted to specific audiences.[44]

In 1998, the American Osteopathic Association launched what became known as the "Unity Campaign." Working with AACOM, state and specialty associations and other groups, the AOA gave renewed attention to the problem of public recognition and understanding of the profession. The intended goal of their efforts was to highlight osteopathic distinctiveness.[45] The AOA taught DOs in the field how to make effective contacts with the media and to develop story lines that health reporters might cultivate. Also, the association unveiled two series of magazine advertisements—both of which were eye-catching and had the potential of effectively conveying to the public the scope of practice of DOs and the wide range of specialties in which they are engaged. However, the annual funding for this national advertising campaign has to date been in the range of only half a million dollars, far too little to buy large and repeated ads in the magazines they have selected and thereby make any appreciable audience impression.[46]

Some younger members of the profession have tried to circumvent the problem of limited funding for public education by contacting producers of prime time network medical series and trying to convince them to prominently feature DO characters in their programs. This effort has yet to be successful. However, in the fall of 2002, a new cable television weekly drama, "Body and Soul" premiered with one of the central members cast as a DO. In addition to incorporating osteopathic philosophy and manipulative medicine, she uses a variety of alternative approaches as well as conventional procedures in her patient management. Though her osteopathic identity has not been a story line during the initial season, this nonetheless appears to have been the first DO character to be incorporated in a U.S. television program.

THE FUTURE

As a profession, osteopathic medicine will face difficult challenges in the near future. It will need to develop increasing financial support for its schools to broaden and strengthen their undergraduate programs. It will have to find ways of creating more good-quality, well-funded internships and residencies under osteopathic auspices. It will need to deal with both governmental and private third party payers over the level of reimbursement and coverage of osteopathic services. It will have to draw back into its orbit many allopathically trained DOs who either have been alienated from the profession or simply see no reason to reestablish their osteopathic ties. The profession will also have to determine in what ways and to what extent it or certain aspects of some of its practitioners' methods constitute "alternative medicine," for the purposes of self-identification and for explaining itself to external audiences.[47] It will have to consider what relationships it should establish or maintain with colleges and societies in more than a dozen countries where there are practitioners who go by the title "osteopaths" and who teach and espouse a drugless, manipulation-based approach to health care.[48] The profession must answer challenges from other health professions—from allopathic medicine, which desires the absorption of DOs into their ranks; from chiropractic, which seeks to legally prohibit any group other than themselves from administering spinal manipulation; from physician assistants and nurse practitioners, who are increasingly duplicating the functions of primary care practitioners and who are seeking the right to practice largely if not completely independently of licensed physicians.[49]

The profession must also consider its rationale for its independent existence. Osteopathic medicine was founded in the late nineteenth century on the idea that a new type of practitioner—the DO—could make a significant difference and improvement in health care. More than 125 years have elapsed since Andrew Taylor Still metaphorically first waved "the banner of osteopathy," yet this initial reason for existence continues to motivate many of those who practice osteopathic medicine. Though the scope of osteopathic teaching and practice has radically changed, the challenge of doing something distinctive and arguably better remains very much a part of the social movement aspect of osteopathic medicine.

Many DOs continue to believe that despite the considerable similarities between themselves and their MD counterparts, there are some essential qualities that make them truly different—whether it be their ability to diagnose and treat through their hands, the quality of their interpersonal skills, their emphasis upon primary care, their commitment to practicing in underserved areas, or their holistic approach to the patient. Some of these perceived differences may be more imagined than real, but the shared beliefs themselves are important to osteopathic identity formation and provide the basis for an ideological rationale for maintaining an independent professional existence.[50]

The strength of the internalized belief that DOs are different and in some respects better than their MD counterparts will likely be dependent upon how the profession is perceived in society. The two prizes that most DOs wish for their profession are, first, convincing scientific evidence that the distinctive aspects of their practice do make a positive difference in patient care and, second, widespread public understanding and appreciation for who they are and what they do. How well and how quickly osteopathic medicine progresses in reaching these goals may determine whether it will exist in the future as an independent and parallel profession of medicine. Failure to make significant progress in these areas may well mean a diminished confidence in the value of maintaining their independence and greater assimilation into the medical mainstream. However, if the osteopathic profession increasingly proves the value of its approach to patient care, and is so recognized by the public, another more remarkable possibility could occur. The potential would exist for osteopathic medicine to not only survive but conceivably begin a new chapter in its history—one that no similar medical movement or parallel profession, such as homeopathy or eclectic medicine, has ever achieved. It could make the leap from being regarded only as a "medical minority" to becoming more broadly recognized as a "medical elite." The DO degree—long viewed as a handicap to public understanding by some practitioners—might instead come to symbolize a special and esteemed type of practitioner, and its use by osteopathic physicians might provide them with a competitive advantage over others in the medical market place.

Realistically, given its current medical and social position, it would likely be a lengthy, difficult, and expensive process to achieve this most desirable status. It would require a large number of committed practitioners and supportive laypersons and significant public and private resources to underwrite and support excellence in osteopathic education, research, and clinical services. Nevertheless, some in this profession have already taken a necessary first step. They have dedicated themselves to furthering core osteopathic beliefs and practices and they are emphasizing to patients and others their distinctiveness from rather than their similarities to other physicians. But many other DOs, particularly younger practitioners, would need to follow that lead and choose to practice distinctively, to engage in research on the fundamental precepts upon which their profession rests, and to fight for their continued autonomy and independence. What course they will pursue is by no means clear. Literally as well as figuratively, the future of osteopathic medicine may ultimately rest in the DOs' own hands—and how they use them.

NOTES

CHAPTER 1. ANDREW TAYLOR STILL

1. Andrew Taylor Still, *The Autobiography of Andrew Taylor Still* (Kirksville, Mo.: by the author, 1908), p. 18.

2. For an extended description and analysis of the circuit system see William Sweet, *Methodism in American History* (Nashville: Abingdon Press, 1954), pp. 143–85.

3. Horace Bushnell quoted in Winthrop S. Hudson, *American Protestantism* (Chicago: University of Chicago Press, 1961), p. 92.

4. Still, *Autobiography*, pp. 18, 19–21.

5. Sweet, *Methodism*, p. 158.

6. Roy Nichols, "The Kansas-Nebraska Act: A Century of Historiography," *Mississippi Valley Historical Review* 48 (1956): 187–212.

7. Still, *Autobiography*, p. 76. Also see Thomas Goodrich, *War to the Knife: Bleeding Kansas, 1854–1861* (Mechanicsburg, Pa.: Stackpole Books, 1998).

8. D. W. Wilder, *The Annals of Kansas* (Topeka: Kansas Publishing House, 1886), pp. 409–11; Still, *Autobiography*, pp. 73–81.

9. Still, *Autobiography*, p. 76.

10. See William Norwood, *Medical Education in the United States before the Civil War* (Philadelphia: University of Pennsylvania Press, 1944) and Martin Kaufman, *American Medical Education: The Formative Years, 1765–1910* (Westport, Conn.: Greenwood Press, 1976). According to Thomas Bonner, the number of self-taught practitioners in Kansas was quite high: "In the heroic early years of Kansas medicine when settlers were scattered and doctors few, a doctor might almost be defined as anyone who was practicing medicine. No questions were asked if the practitioner brought relief from suffering. Frequently rough in dress and speech, half-literate, without formal training in medicine, the amateur physician was nevertheless on occasion a man of talents" (Thomas Bonner, *The Kansas Doctor* [Lawrence: University of Kansas Press, 1959], p. 11).

11. William Rothstein, *American Physicians in the Nineteenth Century: From Sects to Science* (Baltimore: Johns Hopkins University Press, 1972), pp. 85–87.

12. The makeup of A. T. Still's early medical library has not been completely settled. A letter dated October 29, 1943, from Dr. M. D. Warner, then dean of the Kirksville College, to American Osteopathic Association editor Ray Hulburt indicates that the school possessed a number of Still's early books, including an unidentified 1845 dispensatory; Johannes Muller's *Elements of Physiology* (Philadelphia: Lea and Blanchard, 1843); Robert Druitt's *Principles and Practice of Modern Surgery* (Philadelphia: Lea and Blanchard, 1842); William Ferguson's *A System of Practical Surgery* (London: J. Churchill, 1842); and Robert Harrison's *The Dublin Dissector* (New York: S. S. and W. Wood, 1858). In the intervening years many of these books have, unfortunately, disappeared from the school's archives. However, two others have been discovered: Robley Dunglison's *The Practice of Medicine: A Treatise on Special Pathology and Therapeutics*, vol. 2 (Philadelphia: Lea and Blanchard, 1844), and Samuel Cooper's *The First Lines of the Practice of Surgery*, vol. 1 (New York: James and John Harper Printers, 1822).

13. Still, *Autobiography*, pp. 57, 84–85. In his colorful but wholly undocumented biography of his grandfather, Charles Still makes the remarkable and dubious claim that Andrew Still obtained the local Indian chief's permission to disinter his tribe's corpses to further his knowledge of anatomy. See Charles E. Still Jr., *Frontier Doctor, Medical Pioneer: The Life and Times of A.T. Still and His Family* (Kirksville, Mo.: Thomas Jefferson University Press, 1991) p. 23.

14. In a document dated December 17, 1877, pertaining to his application for an army pension, Still declared, "I was surgeon *but* the adjutant placed me on the roll as a hospital steward and I was paid as such. The whole reg[iment] will testify to the truthfulness of the statement. I did the duty of surgeon." (Pension File of Andrew Taylor Still. Chicago, American Osteopathic Association Archives). See also Still, *Autobiography*, p. 186; Andrew Taylor Still, "Dr. Still's Department," *Journal of Osteopathy* (hereafter *J Ost*) 7 (August 1900): 98–99.

15. The article was published in the *Ladies' Home Journal* in 1908 and republished in *Concerning Osteopathy*, ed. George Webster (Carthage, N.Y.: Cruikshank and Ellsworth, 1910), pp. 55–66. After the article appeared, a skeptical reader of the journal the *Osteopathic Physician* requested proof. Its editor, Henry Bunting, wrote for clarification to Still's son-in-law, George Laughlin, who in turn asked A. T. Still. The latter thereupon corrected the details of the story. See "Asks If A.T. Still Ever Was a Real Doctor," *Osteopathic Physician* (hereafter *OP*) 15 (January 1909): 8. For the first medical school established in Kansas City see Carrie Whitney, *Kansas City, Missouri: Its History and People, 1808–1908*, 2 vols. (Chicago: S. J. Clarke Publishing, 1908), 1:475.

16. Benjamin Rush, *An Account of the Bilious Remitting Yellow Fever: As it Appeared in the City of Philadelphia, in the year 1793* (Philadelphia: Thomas Dobson, 1794). See also Saul Jarcho, "John Mitchell, Benjamin Rush, and Yellow Fever," *Bulletin of the History of Medicine* (hereafter *Bull Hist Med*) 31 (1957): 132–36.

17. See Lester King, *Transformations in American Medicine: From Benjamin Rush to William Osler* (Baltimore: Johns Hopkins University Press, 1991) pp. 1–65; Rothstein, *American Physicians*, pp. 45–49. As late as 1912 William Osler was still advocating the use of bloodletting in treating pneumonia. See his *The Principles*

and Practice of Medicine, 8th ed. (New York: D. Appleton, 1912), pp. 99–100. According to Bonner, bloodletting was used quite extensively in Kansas throughout the 1850s, less so in the 1860s, and rarely in the 1870s. Bonner, *Kansas Doctor*, pp. 20–21. For the decline of bloodletting and heroic medicine nationally see John Harley Warner, *The Therapeutic Perspective: Medical Practice, Knowledge, and Identity in America, 1820–1885* (Cambridge: Harvard University Press, 1986).

18. See G. B. Risse, "Calomel and the American Medical Sects during the Nineteenth Century," *Mayo Clinic Proceedings* 48 (1973): 57–64. Also Rothstein, *American Physicians*, pp. 50–60; Bonner, *Kansas Doctor*, p. 21.

19. Risse, "Calomel and American Medical Sects," p. 58.

20. Still, *Autobiography*, p. 298.

21. Erwin Ackerknecht, "Aspects of the History of Therapeutics," *Bull Hist Med* 36 (1962): 400–412.

22. Jacob Bigelow, *A Discourse on Self-Limited Disease* (Boston: Ticknor and Fields, 1854), p. 4.

23. Oliver Wendell Holmes, *Medical Essays, 1842–1882* (Boston: Houghton Mifflin, 1892), p. 203.

24. See Bonnie Blustein, *"Preserve Your Love of Science": Life of William A. Hammond, American Neurologist* (New York: Cambridge University Press, 1991).

25. Bonner, *Kansas Doctor*, pp. 10–47, 23–30.

26. Andrew Taylor Still, "Dr. Still's Department," *J Ost* 6 (August 1899): 92–93. See also Bonner, *Kansas Doctor*, p. 21.

27. Still, *Autobiography*, pp. 87–88.

28. Ibid., p. 88.

29. For overviews of these movements see Norman Gevitz, ed., *Other Healers: Unorthodox Medicine in America* (Baltimore: Johns Hopkins University Press, 1988) and James Whorton, *Nature Cures: The History of Alternative Medicine In America* (New York: Oxford University Press, 2002).

30. See Samuel Thomson, *The New Guide to Health* (Boston: by the author, 1832); John Haller, *The People's Doctors: Samuel Thomson and the American Botanical Movement* (Carbondale, Ill.: Southern Illinois University Press, 2000); Rothstein, *American Physicians*, pp. 125–51; Joseph Kett, *The Formation of the American Medical Profession: The Role of Institutions, 1780–1860* (New Haven: Yale University Press, 1968) pp. 97–131.

31. Kett, *Formation of the American Medical Profession*, p. 23.

32. See Samuel Hahnemann, *Organon of Homeopathic Medicine* (New York: William Radde, 1849). For a critique of his approach in historical context see Norman Gevitz, "Unorthodox Medical Theories," in *Companion Encyclopedia of the History of Medicine*, 2 vols. (London: Routledge, 1993) 1:603–33.

33. See Martin Kaufman, *Homeopathy in America: The Rise and Fall of a Medical Heresy* (Baltimore: Johns Hopkins University Press, 1972); Rothstein, *American Physicians*, pp. 152–73, 230–43, 294–301; For an insightful biography of its leading school see Naomi Rogers, *An Alternative Path: The Making and Remaking of Hahnemann Medical College and Hospital of Philadelphia* (New Brunswick, N.J.: Rutgers University Press, 1998).

34. Holmes, *Medical Essays*, pp. ix–x.

35. Rothstein, *American Physicians*, pp. 237–39.

36. John Haller, *Medical Protestants: The Eclectics in American Medicine, 1825–1939* (Carbondale: Southern Illinois University Press, 1994); Rothstein, *American Physicians*, pp. 217–29. Those botanic practitioners whose views were more similar to those of Thomson formed the smaller medical sect named physio-medicalism. See John Haller, *Kindly Medicine: Physiomedicalism in America, 1836–1911* (Kent, Ohio: Kent State University Press, 1997).

37. See the table of students and graduates by sect, from 1850 to 1920, in Rothstein, *American Physicians*, p. 287.

38. Still, "Dr. Still's Department," *J Ost* 6 (August 1899): 92–93.

39. For the rise of temperance in Methodism, see Sweet, *Methodism*, pp. 171–72, 241–36. For temperance as a social phenomenon, see Joseph Gusfield, *Symbolic Crusade: Status Politics and the American Temperance Movement* (Urbana: University of Illinois Press, 1963), pp. 13–57.

40. Still, *Autobiography*, p. 83.

41. Richard Shryock, "Sylvester Graham and the Popular Health Movement, 1830–1870," *Mississippi Valley Historical Review* 18 (1931): 172–83; 178. For a more comprehensive view of Graham and others see James Whorton, *Crusaders for Fitness: The History of American Health Reformers* (Princeton: Princeton University Press, 1982).

42. Sylvester Graham, *Lectures on the Science of Human Life*, 2 vols. (Boston: Marsh, Capen, Lyon and Webb, 1839).

43. Marshall Scott Legan, "Hydropathy in America: A Nineteenth-Century Panacea," *Bull Hist Med* 45 (1971): 274.

44. For more comprehensive portraits of hydropathy see Susan Cayleff, *"Wash and Be Healed": The Water-Cure Movement and Women's Health* (Philadelphia: Temple University Press, 1986); and Jane Donegan, *"Hydropathic Highway to Health": Women and Water-Cure in Antebellum America* (Westport, Conn.: Greenwood Press, 1986).

45. This story is related on pp. 195–201 of an unpublished, untitled manuscript by his sister Marovia Still Clark which is preserved in the archives of the Kirksville College of Osteopathic Medicine, Kirksville, Missouri.

46. For a brief discussion of those healers who employed stroking earlier, see Vincent Buranelli, *The Wizard from Vienna: Franz Anton Mesmer* (New York: Coward, McCann, and Geoghagen, 1975), pp. 17–25. One contemporary of Mesmer deserving of special mention was the American Elisha Perkins (1741–1799), who employed two metallic rods that he called tractors to perform similar cures. Believing that his discovery would benefit acute, infectious disease as well as nervous disorders, Perkins lost his life testing his theory during a yellow fever epidemic in New York. See Jacques Quen, "Elisha Perkins, Physician, Nostrum-Vendor or Charlatan?" *Bull Hist Med* 37 (1963): 159–66. For the continuing influence of Mesmer see Adam Crabtree, *From Mesmer to Freud: Magnetic Sleep and the Roots of Psychological Healing* (New Haven: Yale University Press, 1993).

47. Buranelli, *Wizard from Vienna*, pp. 59–132.

48. *Report of Dr. Benjamin Franklin and Other Commissioners Charged . . . with the Examination of the Animal Magnetism* (London: printed for L. Johnson, 1785); Buranelli, *Wizard from Vienna*, pp. 157–68.

49. *Report of the Magnetical Experiments Made by the Commission of the Royal Academy of Paris, Read on the Meetings of June 21st and 28th, 1831, by Mr. Husson, the Reporter*, trans. and intro. Charles Poyen (Boston: D. K. Hitchcock, 1836). James Braid, *Neurypnology: Or the Rationale of Nervous Sleep: Considered in Relation with Animal Magnetism* (London: J. Churchill, 1843).

50. Eric Carlson, "Charles Poyen Brings Mesmerism to America," *Journal of the History of Medicine* (hereafter *J Hist Med*) 15 (1960):121–32.

51. Charles Poyen, *Progress of Animal Magnetism in New England* (Boston: Weeks, Jordan, 1837); C. F. Durant, *Exposition; or, A New Theory of Animal Magnetism* (New York: Wiley and Putnam, 1837); W. L. Stone, *Letter to Dr. A. Brigham on Animal Magnetism* (New York: George Dearborne, 1837); A Gentleman from Philadelphia, *The Philosophy of Animal Magnetism* (Philadelphia: Merrihew and Gunn, 1837).

52. For a description of Quimby's practice and differing accounts of his relative impact upon Eddy, see Frank Podmore, *From Mesmer to Christian Science: A Short History of Mental Healing* (New York: University Books, 1963), pp. 250–99; and Robert Peel, *Mary Baker Eddy: The Years of Discovery, 1821–1875* (New York: Holt, Rineheart and Winston, 1966), pp. 146–338. See also Annetta Dresser, *The Philosophy of P.P. Quimby* (Boston: George Ellis, 1895); Mary Baker Eddy, *Science and Health with Key to the Scriptures* (Boston: First Church of Christ Scientist, 1971). For a recent account of ordinary Christian Science practitioners see Rennie Schoepflin, *Christian Science on Trial* (Baltimore: Johns Hopkins University Press, 2003).

53. Robert Delp, "Andrew Jackson Davis: Prophet of American Spiritualism," *Journal of American History* 20 (1967): 43–57. See also S. E. D. Shortt, "Physicians and Psychics: Medical Response to Spiritualism," *J Hist Med* 39 (1984): 339–55; and Edward M. Brown, "Neurology and Spiritualism in the 1870s," *Bull Hist Med* 57 (1983): 563–77.

54. Andrew Jackson Davis, *The Great Harmonia*, 4 vols. (Boston: Benjamin Mussey, 1853) 1:325–30.

55. See John Teahan, "Warren Felt Evans and Mental Healing," *Church History* 48 (1979): 63–80.

56. Warren Felt Evans, *Mental Medicine* (Boston: H. H. Carter, 1885) p. 109.

57. Edwin Dwight Babbitt, *Vital Magnetism* (New York: by the author, 1874). See also James Mack, *Healing By Laying On of Hands* (Boston, Colby and Rich, 1879) pp. 164–171, which contains Babbitt's "Rules for Magnetizers."

58. *Banner of Light* 36 (January 9, 1875): 8. This letter was an appeal for assistance on behalf of the people of Baldwin, who were suffering the effects of a grasshopper invasion. Written by Henry Durgin and co-signed by Still and three others, it noted, "There are a few workers here [spiritualists] but we are looked upon as 'crazy' and 'worse than infidels' and any calamity that may fall upon us is construed to be the just judgment of God, for our daring to think and act for ourselves." See Carol Trowbridge, *Andrew Taylor Still, 1828–1917* (Kirksville, Mo.:

Thomas Jefferson University Press, 1991): 106–11. In 1903 Still attended a spiritualist meeting in Clinton, Iowa, whereupon he reported back to his students that the spiritualists' views on drugging were similar to his own. Fanny Bennett and Ida Bush, "Dr AT Still's Visit to the Spiritualists' Meeting at Clinton, Iowa," *Bulletin of the Axis and Atlas Clubs* 4 (September 1903): 3–6.

59. See in particular Andrew Taylor Still, *Philosophy of Osteopathy* (Kirksville, Mo.: by the author, 1899), pp. 44, 48, 57, 65.

60. Still, *Autobiography*, p. 182. Perhaps this idea stemmed from his training as an orthodox physician, for, as Bonner has noted of prewar Kansas practitioners generally, "From the books they had read and the medical colleges they had attended they viewed disease as largely the work of the blood" (Bonner, *Kansas Doctor*, pp. 21–22).

61. In her biography of Still, Carol Trowbridge attempts to connect the origin of his beliefs to the writings of evolutionary theorists. She believes that Still was directly indebted to Herbert Spencer's *First Principles* (1859) and *Principles of Biology* (1866). However, these two books compared with Still's four published volumes, are quite disparate in terms of epistemological foundations, methods of inquiry, data collection, use of source materials, cosmologic perspective, and subject matter. Also, in the brief passages of Still's works where he alludes to evolutionary themes, Still's thinking is muddled and he appears to have profited little by any of the notable evolutionary theorists he may have read. Furthermore, Spencer's works have nothing remotely to do with palpatory diagnosis and manipulative treatment directed at the spine. See Trowbridge, *Andrew Taylor Still*.

62. Still, *Autobiography*, p. 108.

63. See E. M. Violette, *History of Adair County* (Kirksville, Mo.: Journal Printing, 1911), p. 343. *North Missouri Register*, March 11, 1875. This notice ran in the *Register* until August of that year. From this time forward Still does not appear to have used the newspapers for advertising purposes.

64. Nine years later he registered as a physician and surgeon in Adair County. Photos of his certificates are printed in Arthur Hildreth, *The Lengthening Shadow of Andrew Taylor Still* (Kirksville, Mo.: Journal Printing, 1942), pp. 294–95.

65. Still, *Autobiography*, p. 112. Pension File of Andrew Taylor Still.

66. E. R. Booth, *History of Osteopathy and Twentieth-Century Medical Practice* (Cincinnati: Caxton Press, 1924), pp. 505–6.

67. Dr. Percival Potts said of his famous eighteenth-century bonesetting contemporary "Crazy Sally" Mapp, "Even the absurdities and impracticality of her own promises and engagements, were by no means equal to the expectations or the credulity of those who ran after her, that is, of all ranks and degrees of people from the lowest laborer up to those of the most exalted rank and station, several of whom not only did not hesitate implicitly to believe the most extravagant assertions of this ignorant, drunken, female savage, but even solicited her company or at least seemed to enjoy her society" (Quoted in C. J. S. Thompson, *The Quacks of Old London* [London: Brentano, 1928], p. 303).

68. James Paget, "Cases That Bonesetters Cure," *British Medical Journal* 1 (January 5, 1867): 1–4.

69. Wharton Hood, *On Bonesetting, So-Called, and Its Relation to the Treatment of Joints Crippled by Injury* (London: Macmillan, 1871), pp. 4, 26–27.

70. Ibid., pp. 149, viii.

71. Robert J. T. Joy, "The Natural Bonesetters, with Special Reference to the Sweet Family of Rhode Island," *Bull Hist Med* 28 (1965): 416–41.

72. Douglas Graham, *A Practical Treatise on Massage: Its History, Mode of Application, and Effects* (New York: William Wood, 1884), p. 20.

73. Some bonesetters were apparently active in and around Missouri during this era. One MD wrote of an individual claiming to be one of the Rhode Island Sweets working his trade in his near neighborhood. See A. J. Steele, "The Osteopathic Fad," *Transactions of the Medical Association of the State of Missouri* (1895): 343–58.

74. Still, *Autobiography*, pp. 100–101.

75. Homer Bailey quoted in Booth, *History of Osteopathy*, p. 32. See also Hildreth, *Lengthening Shadow*, p. 382.

76. Examples in Still's approach to public speaking can be found in his *Autobiography*. See pp. 153–66, 177–81.

77. Harry Still quoted in Booth, *History of Osteopathy*, p. 57.

78. J. O. Hatten quoted in ibid., pp. 61–62. For a patient's narrative of Still's work in the 1880s, see the interview with George Compton in the *Kirksville Graphic*, March 19, 1897, reprinted in *J Ost* 3 (March 1897): 7–8.

79. Henry Bunting, "How Osteopathy Got Its First Recognition in Kirksville," *J Ost* 5 (1899): 473–75; Hildreth, *Lengthening Shadow*, pp. 13–15, 371.

80. Hildreth, *Lengthening Shadow*, p. 26.

81. Andrew Taylor Still, "Dr. Still's Department," *J Ost* 8 (1901): 68.

CHAPTER 2. THE MISSOURI MECCA

1. See E. R. N. Grigg, "Peripatetic Pioneer: William Smith, MD, DO (1862–1912)," *J Hist Med* 22 (1967): 169–79; "Dr. William Smith, Pioneer at His Old Post," *OP* 12 (July 1907): 15–16.

2. William Smith as quoted in E[amons] R. Booth, *History of Osteopathy and Twentieth-Century Medical Practice* (Cincinnati: Caxton Press, 1924), p. 448.

3. Arthur Hildreth, *The Lengthening Shadow of Andrew Taylor Still* (Kirksville, Mo.: Journal Printing, 1942), p. 31. The problem of securing sufficient material for dissection was an acute one for the remainder of the century. In 1897, Smith and his assistant Clarence Rider traveled to Chicago, where they were able to obtain unclaimed bodies by paying off attendants at the Cook County morgue. When the "heist" had been discovered and the participants named, a warrant for Smith's arrest was issued. Fortunately, for Still's anatomy instructor, the governor of Missouri refused to authorize his extradition and the matter was subsequently dropped. See "Dr. Clarence Rider Took Part in Stirring Pioneer Affairs," *OP* 8 (November 1905): 13.

4. For discussion of the size of the initial class, see Georgia Walter, *The First School of Osteopathic Medicine* (Kirksville, Mo.: Thomas Jefferson University Press,

1992), pp. 3–11. For an astute evaluation of Smith see Henry Bunting, "Dr. William Smith Died of Pneumonia in Scotland Feb. 15th after Two Days Illness," *OP* 21 (March 1912): 1.

5. Henry Bunting, "The Real A.T. Still," *OP* 32 (December 1917): 16.

6. Andrew Taylor Still, *The Autobiography of Andrew Taylor Still*, (Kirksville, Mo.: by the author, 1908), pp. 184–85.

7. Ibid. Still was fascinated by machinery, was an inventor, and repeatedly used his mechanical experience to illustrate his medical ideas. He had won a local prize in Baldwin for an improved butter churn, reputedly made a significant improvement to a reaper for which he received no credit, and as late as 1910 when he was 82 years old, was issued a U.S. patent for a "smokeless furnace." A copy of this patent is printed in Charles Still, *Frontier Doctor, Medical Pioneer: The Life and Times of A.T. Still and His Family* (Kirksville, Mo.: Thomas Jefferson University Press, 1991): 232.

8. Hildreth, *Lengthening Shadow*, pp. 31–32, 194.

9. Andrew Taylor Still, "To Patients and Visitors," *J Ost* 1 (August 1894): 2.

10. Andrew Taylor Still, "Important to Patients," *J Ost* 2 (December 1895): 7; idem, "Address," *J Ost* 1 (May 1894): 1; idem, "A Plea for Temperance," *J Ost* 1 (June 1894): 1; idem, *Autobiography*, p. 276.

11. *St. Louis Globe-Democrat*, January 16, 1895, reprinted in *J Ost* 1 (January 1895): 8; *Des Moines Daily News*, November 10, 1895, reprinted in *J Ost* 2 (December 1895): 1; *Nebraska Daily Call*, November 7, 1895, reprinted in *J Ost* 2 (December 1895): 4; *Ottumwa (Iowa) Press*, n.d., reprinted in *J Ost* 1 (October 1894): 4; *Bethany (Ill.) Echo*, n.d., reprinted in *J Ost* 1 (February 1895): 2.

12. "Volume Three," *J Ost* 3 (June 1896): 4.

13. *Kirksville Graphic*, n.d., reprinted in *J Ost* 2 (September 1895): 2.

14. *Loyal Workman* (Ottumwa, Ia.), September 1, 1896, reprinted in *J Ost* 3 (October 1896): 1.

15. "A Glance Backward," *J Ost* 4 (1898): 367–74, 384–86.

16. "Invalids from Twenty-One States: A Patient Tells What He Saw at the Still Infirmary in Kirksville," *J Ost* 3 (June 1896): 6. See also the article from *Godey's Magazine* (October 1895) reprinted in *J Ost* 2 (October 1895): 3.

17. These cases were originally published in the *Kirksville Journal*, January 30, 1896, and reprinted in the *J Ost* 2 (February 1896): 3.

18. Cincinnati Commercial Tribune, September 25, 1896, reprinted in *J Ost* 3 (October 1896): 5. See also Hildreth, *Lengthening Shadow*, pp. 85, 129; Booth, *History of Osteopathy*, pp. 46, 99. Senator Foraker's son died at the age of thirty-nine. See "Death of A. St. C. Foraker Recalls Early Osteopathic History," *Forum of Osteopathy* (hereafter *Forum of Ost*) 5 (1931): 176.

19. P. F. Greenwood, "The Position Osteopathy Occupies under the Present State Law," *J Ost* 1 (May 1894): 2.

20. "A Petition," *J Ost* 1 (May 1894): 4.

21. Greenwood, "Position Osteopathy Occupies," p. 2

22. A. J. Steele, "The Osteopathic Fad," *Transactions of the Medical Association of the State of Missouri* (1895): 363.

23. "Sequel to 'The Osteopathic Fad,'" *J Ost* 3 (October 1896): 7.

24. William Smith, "Four Years Ago," *J Ost* 3 (September 1896): 6.

25. Steele, "Osteopathic Fad," p. 356.

26. E. C. Pickler and C. M. T. Hulett quoted in Booth, *History of Osteopathy*, pp. 75, 493.

27. Excerpts from the first charter can be found in the newspaper article reprinted in Hildreth, *Lengthening Shadow*, p. 34. See also Minutes of the Board of Trustees of the American School of Osteopathy, 1892, mimeographed, American Osteopathic Association Archives, Chicago.

28. Andrew Taylor Still, "Editorial," *J Ost* 3 (December 1896): 4.

29. See Booth, *History of Osteopathy*, p. 85.

30. "Legislative," *J Ost* 3 (February 1897): 4; "Missouri in Line," *J Ost* 3 (March 1897): 1.

31. "Missouri in Line," p. 1.

32. Ibid.

33. According to figures cited by E. M. Violette, the number of graduates per year jumped from 48 in 1897, to 136 in 1898, 185 in 1899, and 317 in 1900. See his *History of Adair County* (Kirksville: Journal Printing, 1911), p. 264.

34. This is most evident in the books they wrote, which were among the first texts of the school. See William Smith, *Notes on Anatomy* (Kirksville, Mo.: by the author, 1898); Carl McConnell, *Notes on Osteopathic Therapeutics* (Kirksville: by the author, 1898); Charles Hazzard, *Principles of Osteopathy*, 3rd ed. (Kirksville: Journal Printing, 1899); C. W. Proctor, *A Brief Course in General Chemistry* (Kirksville: by the author, 1898); idem, *A Brief Course in Physiological Chemistry* (Kirksville: by the author, 1898); and J. Martin Littlejohn, *Physiology: Exhaustive and Practical*, 2 vols. (Kirksville: Journal Printing, 1898).

35. See Richard Cyriax, "A Short History of Mechano-Therapeutics in Europe until the Time of Ling," *Janus* 19 (1914): 178–88; as quoted, p. 183; pp. 189–204; as quoted, p. 225. For a popular history see Robert Calvert, *The History of Massage: An Illustrated History From Around the World* (Rochester Vt.: Healing Arts Press, 2002).

36. William Balfour, *Illustrations of the Power of Compression and Percussion in the Cure of Rheumatism, Gout, and Debility of the Extremities in Promoting Health and Longevity* (Edinburgh: P. Hill, 1819); John Bacot, *Observations on the Use and Abuse of Friction with Some Remarks on Motion and Rest, as Applicable to the Cure of Various Surgical Diseases* (London: Callow and Wilson, 1822); William Cleoburey, *A Full Account of the System of Friction as Adopted by J. Grosvenor and Pursued with the Greatest Success in Cases of Contracted Joints and Lameness from Various Causes* (Oxford: Munday and Slatter, 1825); S. Weir Mitchell, *Injuries of Nerves and Their Consequences* (Philadelphia: J. B. Lippincott, 1872); idem, *Fat and Blood: An Essay on the Treatment of Certain Forms of Neurasthenia and Hysteria*, 8th ed. (Philadelphia: J. B. Lippincott, 1900). Mitchell noted, "It is many years since I first saw in this city general massage used by a charlatan in a case of progressive paralysis. The temporary results he obtained were so remarkable that I began soon after to employ it in locomotor ataxia, in which it sometimes proved a signal value, and in other forms of spinal and local disease" (*Fat and Blood*, p. 81).

37. William Murrell, *Massage as a Mode of Treatment* (Philadelphia: P. Blakiston, Son, 1886), pp. 38–40.

38. Douglas Graham, *A Practical Treatise on Massage: Its History, Mode of Application, and Effects* (New York: William Wood, 1884).

39. George Taylor, *Massage: Principles and Practice of Remedial Treatment by Imparted Motion* (New York: Fowler and Wells, 1884), p. 28.

40. Graham, *Practical Treatise on Massage*, p. 35.

41. See Edgar Cyriax, "Concerning the Early Literature on Ling's *Medical Gymnastics*," *Janus* 30 (1926): 225–32; idem, *Bibliographia Gymnastica Medica* (n.p., 1909).

42. Hazzard, *Principles of Osteopathy*, p. 292.

43. See Murrell, *Massage as Mode of Treatment*, pp. 20–28.

44. See Francis Schiller, "Spinal Irritation and Osteopathy," *Bull Hist Med* 45 (1971): 252–54.

45. See in particular Evans Riadore, *A Treatise on the Irritation of the Spinal Nerves as the Source of Nervousness, Indigestion, Functional and Organic Derangements of the Principal Organs of the Body, and on the Modifying Influence of Temperament and Habits of Man over Diseases and Their Importance as Regards Conducting Successfully the Treatment of the Latter; and on the Therapeutic Use of Water* (London: J. Churchill, 1843); and William Griffin and David Griffin, *Observations on the Functional Affections of the Spinal Cord and Ganglionic System of Nerves in which Their Identity with Sympathetic, Nervouse and Irritative Diseases is Illustrated* (London: Burgess and Hill, 1844).

46. John Hilton, *On Rest and Pain: A Course of Lectures on the Influence of Mechanical and Physiological Rest on the Treatment of Accidents and Surgical Diseases, and the Diagnostic Value of Pain*, 2nd ed. (New York: W. Wood, 1879).

47. Francis Schiller, noting the considerable number of works published on the subject of spinal irritation during the second and third quarters of the nineteenth century, has suggested that Still could hardly have been unaware of this doctrine. Schiller, however, was unable to marshal any hard evidence that spinal irritation had a direct influence upon Still's theory, nor have I subsequently been able to uncover any new information that would show an incontrovertible link between them. See Schiller, "Spinal Irritation and Osteopathy," pp. 250–66.

48. Hazzard, *Principles of Osteopathy*, pp. 8–13; J. Martin Littlejohn, *Principles of Osteopathy* (Chicago: by the author, 1902), pp. 2–8.

49. Walter Riese, *A History of Neurology* (New York; MD Publications, 1959), pp. 131–36. See also J. M. D. Olmstead, *Charles-Edouard Brown-Sequard: A Nineteenth Century Neurologist and Endocrinologist* (Baltimore: Johns Hopkins Press, 1946).

50. Hazzard, *Principles of Osteopathy*, pp. 35–51.

51. See William Bulloch, *The History of Bacteriology* (New York: Dover Publications, 1979).

52. Andrew Taylor Still, "Smallpox," *Bulletin of the Atlas and Axis Clubs*, no. 3 (1901): pp. 6–7.

53. See James Littlejohn, "Bacteriology: Its History and Relation to Disease,"

J Ost 5 (1898): 130–34; David Littlejohn, "Diseases of a Pathogenic Origin: Indications for Treatment from an Osteopathic Standpoint," *J Ost* 5 (1898): 177–80; and Carl McConnell, *The Practice of Osteopathy* (Chicago: The Hammond Press, 1899).

54. For example, see Andrew Taylor Still, *Philosophy of Osteopathy* (Kirksville, Mo.: by the author, 1899), p. 12.

55. For more on the American School of Osteopathy see Walter, *The First School of Osteopathic Medicine*, pp. 1–147.

CHAPTER 3. IN THE FIELD

1. "Graduates of the American School of Osteopathy," *J Ost* 7 (1900): 244–48.

2. Andrew Taylor Still, "Dr. Still's Department," *J Ost* 8 (1901): 68.

3. "What is the Science of Osteopathy?" *Cosmopolitan Osteopath* 1 (November 1898): 10.

4. Therese Cluett, "The Amusing Side of Osteopathy," *Boston Osteopath* 3 (May 1900): 90.

5. Herbert Bernard quoted in E[amons] R. Booth, *History of Osteopathy and Twentieth-Century Medical Practice* (Cincinnati: Caxton Press, 1924), p. 60.

6. "Why Should I Try Osteopathy?" *Southern Journal of Osteopathy* 1 (November 1898): 325.

7. "Notice to Our Patrons," *Northern Osteopath* 1 (March 1897): 6.

8. "Diseases Successfully Treated by Osteopathy," *Pennsylvania Journal of Osteopathy* 1 (June 1899): 24.

9. "Diseases Treated," *New York Osteopath* 1 (April 1898) 47.

10. "Testimonials: Why We Run Them," *Osteopathic Success* 1 (February 1901): 16.

11. "Miscellaneous Cases Reported from the Field," *Popular Osteopath* 1 (1899): 156.

12. W. L. Riggs, "Osteopathy in Acute Diseases," *Cosmopolitan Osteopath* 2 (July 1899): 23.

13. A. L. Evans, "Quick Cures," *Popular Osteopath* 1 (1899): 150–51.

14. Cluett, "Amusing Side of Osteopathy," p. 90.

15. "Letters from Graduates," *J Ost* 4 (1898): 444.

16. Philadelphia College and Infirmary of Osteopathy, *Annual Announcement* (1899), p. 16.

17. A. L. Evans, "Why Osteopathy is Popular," *Popular Osteopath* 1 (January 1899): 15.

18. See "Is Osteopathy Dying Out? Have We Passed Our Zenith?" *OP* 14 (November 1908): 1.

19. See Booth, *History of Osteopathy*, pp. 170–71; Charles Still, "Establishing the Fact that Osteopathy Is a Science," *J Ost* 4 (1898): 415–18.

20. "The Man Who Took Osteopathy to the Pacific," *OP* 8 (June 1905): 13.

21. Booth, *History of Osteopathy*, pp. 179–80, 181–83.

22. Ibid., p. 193.

23. See discussion in ibid., pp. 162–201.

24. Ibid., pp. 191–92.

25. Ibid., pp. 106–7; Arthur Hildreth, The *Lengthening Shadow of Andrew Taylor Still* (Kirksville, Mo.: Journal Printing, 1942), p. 24.

26. As reprinted in W. Livingston Harlan, *Osteopathy: The New Science* (Chicago: by the author, 1898), pp. 58–59.

27. "North Dakota Grit," *J Ost* 3 (February 1897): 1–2.

28. Albert Bigelow Paine, ed., *Mark Twain's Speeches* (New York: Harper and Brothers, 1923) p. 233.

29. Albert Bigelow Paine, *Mark Twain, A Biography*, 3 vols. (New York: Harper and Brothers, 1912) 2:1087–88. For more on Twain and Osteopathy see M. M. Brashear, "Dr. Still and Mark Twain," *Journal of the American Osteopathic Association* (hereafter *JAOA*) 73 (1973): 67–71; and Carol Trowbridge, *Andrew Taylor Still: 1828–1917* (Kirksville, Mo.: Thomas Jefferson University Press, 1991), pp. 189–93.

30. See Booth, *History of Osteopathy*, pp. 95–161.

31. Ibid., p. 163.

32. For a history of this process see William Rothstein, *American Physicians in the Nineteenth Century: From Sects to Science* (Baltimore: Johns Hopkins University Press, 1972), pp. 298–326; Martin Kaufman, *Homeopathy in America: The Rise and Fall of a Medical Heresy* (Baltimore: Johns Hopkins University Press, 1972), pp. 141–73. For a comparison of these sects' colleges with osteopathic schools after the turn of the century see Norman Gevitz, "The Fate of Sectarian Medical Education," in *Beyond Flexner: Medical Education in the Twentieth Century*, ed. Barbara Barzansky and Norman Gevitz (Westport, Conn.: Greenwood Press, 1992), pp. 83–97.

33. These estimates are based on an examination of early alumni lists published by the colleges as well as statistics cited in Booth, *History of Osteopathy*, pp. 71–94.

34. Ibid., p. 87.

35. See, for example, the advertisement for the American College of Osteopathic Medicine and Surgery on the back cover of the *Journal of the Science of Osteopathy* (hereafter *J Sci Ost*) 1 (December 1900).

36. For a contemporary critique of osteopathic colleges see C. M. Turner Hulett, "The Profession and the Schools," *J Sci Ost* 1 (June 1900): 141–44.

37. "Osteopathy as a Profession," *Cosmopolitan Osteopath* 1 (July 1898): 34.

38. Mason Pressly, "Osteopathy as a Business," *Northern Osteopath* 2 (May 1898): 7–8.

39. "The Advantages of Osteopathy as a Study and a Profession," *Philadelphia Journal of Osteopathy* 1 (January 1899): 13.

40. "The Atlantic School of Osteopathy," *Pennsylvania Journal of Osteopathy* 1 (June 1899): 23.

41. "The Science of Osteopathy," *The Osteopath* 1 (February 1897): 4.

42. This estimate is based upon available early alumni records of several of the colleges. Of 765 total graduates to date listed by the American School of Osteopathy in 1900, for example, 183 (23.9 percent) were women. See "Graduates

of the American School of Osteopathy," pp. 244–48. See also Mike Fitzgerald, "Women in History: Pioneers of the Profession," *The DO* 25 (1984): 67–71. For the struggle of women entering orthodox medical education see Mary Roth Walsh, *"Doctors Wanted, No Women Need Apply": Sexual Barriers in the Medical Profession, 1835–1975* (New Haven: Yale University Press, 1977).

43. Lawrence Finn, "The Location of the Southern School and Infirmary," *Southern Journal of Osteopathy* 1 (February 1898): 13.

44. "Editorial," *The Osteopath* 2 (August 1898): 13–14.

45. "Advantages of Des Moines—Disadvantages of Kirksville," *Cosmopolitan Osteopath* 1 (August 1898): 36–37.

46. Elmer Barber, *Osteopathy: The New Science of Healing* (Kansas City: Hudson-Kimberly Publishing, 1896). Also see his *Osteopathy Complete* (Kansas City: Hudson-Kimberly Publishing, 1898).

47. When Smith temporarily left the American School of Osteopathy for private practice after completion of the first class, his teaching duties were taken over by Jenette "Nettie" Bolles (died 1930), one of his students who had previously earned two bachelor's degrees. The third class in anatomy was conducted by Summerfield Still (1851–1931), the founder's nephew. Each was to found a school; the former with her husband established the Bolles Institute of Osteopathy in Denver, and the latter set up the S. S. Still College of Osteopathy in Des Moines.

48. For details of the Kansas City case, see Booth, *History of Osteopathy*, pp. 86–87, 166–67.

49. See E. M. Violette, *History of Adair County* (Kirksville, Mo.: Journal Printing, 1911), pp. 274–75.

50. For details see "Columbian School of Osteopathy, Medicine and Surgery," *Columbian Osteopath* 2 (October 1899): 263, 265; Violette, *History of Adair County*, p. 275.

51. J. R. Musick, "Is Osteopathy of Greek Origin?" *J Ost* 5 (1898): 221–25.

52. Andrew Taylor Still, "Dr. Still's Department," *J Ost* 5 (1898): 167, idem, "Medical Osteopathy," *J Ost* 8 (1901): 166.

53. "Cofounder of First Osteopathic College Dies," *Western Osteopath* 23 (1929): 19.

CHAPTER 4. STRUCTURE AND FUNCTION

1. C. M. Turner Hulett, "Historical Sketch of the AAAO," *JAOA* 1 (1901): 1–6; "Proceedings of the Fifth Annual Meeting of the American Association for the Advancement of Osteopathy," *JAOA* 1 (1901): 6–15.

2. "Constitution of the American Osteopathic Association," *JAOA* 1 (1901): 16–17.

3. "Constitution and By-Laws," *JAOA* 9 (1909): 37–43. See also "Should the By-Laws Be Changed?" *JAOA* 11 (1911): 667–71.

4. "Preliminary Report of the A.O.A. Committee on By-Laws," *JAOA* 18 (1919): 304–8; W. A. Gravett, "The New AOA," *JAOA* 19 (1920): 191–93.

5. E[amons] R. Booth, *History of Osteopathy and Twentieth-Century Medical Practice* (Cincinnati: Caxton Press, 1924), pp. 106–8; A. L. Evans, "Legal Status of Osteopathy in the Various States," *JAOA* 2 (1903): 145–47.

6. Chester Cole, "Iowa's Medical Board and Osteopathy Law," *Cosmopolitan Osteopath* 2 (March 1899): 3–4; "Osteopathic Victory in Iowa," *JAOA* 1 (1902): 162–63.

7. See A. G. Hildreth, "Report of the Committee on Legislation with Bill Appended," *JAOA* 5 (1905): 71–75; idem, "Osteopathic Legislation," *JAOA* 2 (1903): 143–44.

8. Samuel Baker, "Physician Licensure Laws in the United States, 1865–1915," *J Hist Med* 39 (1984): 173–97. See also William Rothstein, *American Physicians in the Nineteenth Century: From Sects to Science* (Baltimore: Johns Hopkins University Press, 1972), pp. 305–10.

9. A. G. Hildreth, "Osteopathic Legislation," *JAOA* 4 (1905): 191.

10. American Osteopathic Association, *Yearbook and Directory* (Chicago, 1913), pp. 97–108; American Osteopathic Association, *Yearbook and Directory* (Chicago, 1923), pp. 169–89.

11. Hulett, "Historical Sketch of the AAAO," p. 2.

12. Booth, *History of Osteopathy*, pp. 272–77.

13. "Constitution of the American Osteopathic Association," p. 16.

14. Wilfred Harris, "The Three Year Course: Some Questions for the Profession to Decide," *JAOA* 3 (1904): 373.

15. "Report of the AOA Committee on Education," *JAOA* 2 (1902): 10–19.

16. C. M. Turner Hulett, "The Profession and the Schools," *J Sci Ost* 1 (June 1900): 144.

17. Martin Littlejohn, "The Standard of Education," *JAOA* 1 (1902): 191–93. For more on the Littlejohn brothers and the Chicago school see Theodore Berchtold, *To Teach, To Heal, To Serve* (Chicago: Chicago College of Osteopathic Medicine, 1975).

18. Eamons Booth, "Report of Inspector of Osteopathic Schools," *JAOA* 3 (1903) supplement: 9–20.

19. "Proceedings of the Eighth Annual Meeting of the American Osteopathic Association," *JAOA* 4 (1904): 38, 51.

20. Philadelphia College and Infirmary of Osteopathy and the Osteopathic Hospital of Philadelphia, *Annual Announcement* (1919), p. 11.

21. Chicago College of Osteopathy, *Annual Catalog* (1913), pp. 15–16.

22. "Proceedings of the Philadelphia Meeting," *JAOA* 13 (1914): 727.

23. Henry Bunting, "Let Us Discuss Our Failures with Each Other," *OP* (July 1902): 1–2.

24. "Case Reports," *JAOA* 4 (1904): 96–98, 100–101.

25. A textbook based heavily upon these supplements was later issued. See Carl McConnell, ed., *Clinical Osteopathy* (Chicago: A. T. Still Research Institute, 1917).

26. See Edythe Ashmore, ed., "Case Reports," *JAOA* 3 (February 1904) supplement: 1–40. In the same journal: 3 (June 1904) supplement: 1–34.; 4 (March 1905) supplement: 1–36; 4 (August 1905) supplement: 1–32; 5 (July 1906) sup-

plement: 1–40; 6 (June 1907) supplement: 1–48; 7 (September 1907) supplement: 1–45; 7 (June 1908) supplement: 1–47; 7 (August 1908) supplement: 1–48; 8 (June 1909) supplement: 1–53; 8 (July 1909) supplement: 1–48; 8 (August 1909) supplement: 1–46.

27. See Fred Bischoff and Ray Hulbert, "The A. T. Still Research Institute: An Historical Sketch and A Look Ahead," *JAOA* 25 (1926): 376–77.

28. See John Deason and L. G. Robb, "On the Pathways for the Bulbar Respiratory Impulses in the Spinal Cord," *American Journal of Physiology* 28 (1911): 57–63.

29. See John Deason and associates, *Research in Osteopathy* (Chicago: A. T. Still Research Institute, 1916).

30. Barbara Peterson, "Louisa Burns, DO: Pioneer Researcher," *The DO* 18 (July 1978): 21–22.

31. A summary of Burns's work is to be found in Louisa Burns and associates, *Pathogenesis of Visceral Disease following Vertebral Lesions* (Chicago: American Osteopathic Association, 1948).

32. For a more favorable assessment of Burns's research by a collaborator late in her career see Wilbur V. Cole, "Louisa Burns Memorial Lecture," *JAOA* 69 (1970): 1005–17.

33. "Improper Advertising," *JAOA* 2 (1902): 85–86; *JAOA* 3 (1903): 58–59.

34. "Code of Ethics of the American Osteopathic Association," *JAOA* 4 (1904): 92–96; 95.

35. "Osteopaths Frame Code of Ethics," *OP* 3 (December 1902): 1–2.

36. "Code of Ethics of the American Osteopathic Association," p. 95.

37. "Report of the Committee on Education to the Board of Trustees of the American Osteopathic Association," *JAOA* 7 (1907): 87.

38. Henry Bunting, *The Elementary Laws of Advertising* (Chicago: Novelty News Press, 1914).

39. Data derived from American Osteopathic Association, *Yearbook and Directory* (Chicago, 1918); American Osteopathic Association, *Yearbook and Directory* (Chicago: 1930).

40. "Hard to Distinguish Wolves When They Break into the Fold," *OP* 2 (July 1902): 1; "What We Do to Fakers in New York," *OP* 5 (May 1904): 12.

41. S. C. Matthews, "The Fake Osteopath," *The American Osteopath* 2 (September 1900): 9. See also Joseph Sullivan, "Retrospective," *The American Osteopath* 1 (June 1900): 216; W. L. Riggs, "Opposition to the Growth of Osteopathy," *Cosmopolitan Osteopath* 3 (September 1899): 3.

42. C. M. Turner Hulett, "Correspondence Schools," *JAOA* 1 (1902): 149.

43. "Report of the Committee on Education to the Board of Trustees of the American Osteopathic Association," *JAOA* 7 (1907): 87.

44. Chittenden Turner, *The Rise of Chiropractic* (Los Angeles: Powell Publishing, 1931), pp. 11–16.

45. Ibid., pp. 26–34; Ralph Lee Smith, *At Your Own Risk: The Case against Chiropractic* (New York: Trident Press, 1969), pp. 1–13.

46. Smith, *At Your Own Risk*, pp. 8–11; Turner, *Rise of Chiropractic*, pp. 35–45.

47. Edythe Ashmore, "An Imitation and Its Lessons," *JAOA* 7 (1908): 209–11, 310–11.

48. Turner, *Rise of Chiropractic*, pp. 95, 294.

49. For recent comprehensive histories of chiropractic see J. Stuart Moore, *Chiropractic in America: The History of a Medical Alternative* (Baltimore: Johns Hopkins University Press, 1993); and Walter Wardwell, *Chiropractic: History and Evolution of a New Profession* (St. Louis: Mosby Year Book, 1992).

CHAPTER 5. EXPANDING THE SCOPE

1. Minutes of the Board of Trustees of the American School of Osteopathy, 1892; mimeographed, American Osteopathic Association Archives, Chicago.

2. Andrew Taylor Still, "Dr. Still's Department," *J Ost* 8 (1901): 67.

3. Completed American Osteopathic Association School Questionnaires for the academic year 1903–4, microfilmed, American Osteopathic Association Archives, Chicago.

4. Data derived from American School of Osteopathy, *Annual Catalog* (1908), pp. 24–26; American School of Osteopathy, *Annual Catalog* (1918), pp. 66–68; Los Angeles College of Osteopathy, *Annual Catalog* (1908), p. 30; College of Osteopathic Physicians and Surgeons, *Annual Catalog* (1918), pp. 17–18; Philadelphia College and Infirmary of Osteopathy and the Osteopathic Hospital of Philadelphia, *Annual Announcement* (1918), pp. 21–25; Littlejohn College and Hospital, *Bulletin and Journal of Health, Announcement Number* (1909), p. 15; Chicago College of Osteopathy, *Annual Catalog* (1918), p. 19.

5. See Raymond Ward, "Why Some Osteopaths Study at Medical Colleges," *OP* 31 (January 1917): 25–27.

6. D. V. Moore, "Obstetrics and the General Practitioner," *JAOA* 16 (1917): 1197–98.

7. Central College of Osteopathy, *Annual Announcement* (1906), p. 16.

8. Lillian Whiting, "Can the Length of Labor Be Shortened by Osteopathic Treatment?" *JAOA* 11 (1912): 917–21.

9. Harry Collins, "Origin and Progress of Osteopathic Surgery," *JAOA* 23 (1924): 715–16.

10. George Still, "Advantages and Necessity of Osteopathic Post-Operative Treatment," *JAOA* 18 (1919): 485.

11. See Frank Young, *Surgery from an Osteopathic Standpoint* (Kirksville, Mo.: Journal Printing, 1904); S. L. Taylor, "Borderline Cases between Surgery and Osteopathy," *JAOA* 12 (1912): 148–54; James Littlejohn, "Indications for Surgical Interference in Gynecological Cases," *JAOA* 12 (1913): 331–36; "Dr. George A. Still Calls Case Reports the Profession's Most Vital Problem," *OP* 24 (November 1913): 3–5.

12. [Andrew Taylor Still], "Our Platform," *J Ost* 9 (October 1902): 342, later reprinted in Andrew Taylor Still, *Osteopathy: Research and Practice* (Kirksville, Mo.: by the author, 1910), pp. 4–5.

13. Dain Tasker, "The 'Lesion' Osteopath is Too Narrow," *OP* 3 (January 1903): 3.

14. "'Lesions' and 'Adjuncts': The Discussion Which Occurred at Cleveland on This Subject," *JAOA* 3 (1904): 280, 283.

15. Ibid., pp. 285–88.

16. Carl McConnell, *The Practice of Osteopathy* (Chicago: Hammond Press, 1899); idem, "A Few Thoughts for the Osteopathic Practitioner," *J Ost* (January 1901): 9.

17. See C. W. Young, "Digest of Answers to Twelve Questions," *JAOA* 8 (1909): 358–60, 435–38. This was a survey of 260 DOs on the subjects of diet, water, and "thought direction," as well as how broad osteopathy should be.

18. William Rothstein, *American Physicians in the Nineteenth Century: From Sects to Science* (Baltimore: Johns Hopkins University Press, 1972), pp. 177–97.

19. See Merck and Company, *Manual of Therapeutics and Materia Medica: A Source of Ready Reference for the Physician* (Rahway, N.J., 1899).

20. See Hubert Lechevalier and Morris Solotorovsky, *Three Centuries of Microbiology* (New York: Dover Books, 1974).

21. In a commentary upon the smallpox vaccination, Still declared, "I would not antagonize the popular belief in the efficacy of vaccination but do most emphatically combat the insertion into the human body of putrid flesh of any animal" (Still, *Osteopathy: Research and Practice*, p. 456).

22. Ibid., pp. 300–301, 433–35, 470–72.

23. [Still], "Our Platform," p. 342. There was one exception, however. Still believed that a periodic application of a cantharidin blister to the arm would serve as a safer and more effective preventative against smallpox than vaccination: "My theory is, that the first active occupant of the body by an infectious fever will drive off others and hold possession of the body until its power is spent" (Andrew Taylor Still, "Smallpox: Cantharidin as a Germifuge," *J Ost* 9 [February 1902]: 69).

24. [Andrew Taylor Still], "What Will Become of the MD DO?" *J Ost* 10 (November 1903): 366.

25. W. A. Hinckle, "A Protest against Intellectual Tyranny," *OP* 7 (March 1905): 11–12.

26. Some DOs denied the validity of such studies. Said Dr. Riley Moore, "We as a profession have a few who, trusting blindly in claims made by medical authors and not knowing that medical statistics can be juggled to prove anything profitable, believe in such practice [vaccination]. But it seems to me that those who do have failed to grasp the proper conception of osteopathic principles" (Riley Moore, "Vaccination: Osteopaths Ought to Read up on It," *OP* 13 [January 1908]: 4).

27. "Correspondence," *JAOA* 8 (1908): 90–91.

28. Charles Teall, "Report of the Inspector of Schools," *JAOA* 6 (October 1906) supplement: 18–25.

29. E. S. Comstock, "The Littlejohn College Idea," *JAOA* 11 (1911): 675–76; "A Letter from Dr. Littlejohn," *JAOA* 11 (1911): 727–29.

30. A. B. Shaw, "The California Law and Its Workings," *JAOA* 13 (1914): 283–

84; "Report of the Committee on Education," *JAOA* 12 (1913): 746–47; Dain Tasker, "Notes on Medical Legislation in California," *JAOA* 15 (1916): 398–404.

31. Teall, "Report of the Inspector," pp. 19–20.

32. Des Moines Still College of Osteopathy, *Annual Catalog* (1908), p. 50.

33. American School of Osteopathy, *Annual Catalog* (1911), p. 35.

34. "The Message from Philadelphia," *JAOA* 13 (1914): 720–24.

35. "New Legislation in Oregon Causes Dissension," *OP* 26 (March 1915): 8–9.

36. Henry Bunting, "What is *Materia Medica* Anyway? How Far Are We Against It?" *OP* 27 (June 1915): 5–6.

37. "Sentiment in Oregon Divided on New Law," *OP* 28 (July 1915): 9.

38. As reprinted in E[amons] R. Booth, *History of Osteopathy and Twentieth-Century Medical Practice* (Cincinnati: Caxton Press, 1924), p. 442.

39. "Espousing Academic Freedom Was the Most Notable Work of the Portland Convention," *OP* 28 (August 1915): 1–3.

40. Alfred Crosby, Jr., *Epidemic and Peace* (Westport, Conn.: Greenwood Press, 1976), pp. 206–7. See also A. A. Hoehling, *The Great Epidemic* (Boston: Little, Brown, 1961); and Gina Kolata, *Flu: The Story of the Great Influenza Pandemic and the Search for the Virus That Caused It* (New York: Farrar, Strauss and Giroux, 1999).

41. William Osler, *Principles and Practice of Medicine*, 8th ed. (New York: D. Appleton, 1912), p. 119.

42. Carl McConnell, "Editorial: The Treatment of Influenza," *JAOA* 18 (1918): 83–85; C. C. Reed, "Prevention and Treatment of Influenza," *JAOA* 18 1 (1919): 209–11; L. K. Tuttle and Robert Rogers, "Influenza and Pneumonia Treatment," *JAOA* 18 (1919): 211–14; George McCole, "Spanish or Epidemic Influenza from the Treatment Side," *OP* 35 (June 1919): 1–6.

43. Norman Gevitz, "The Sword and the Scalpel: The Osteopathic War to Enter the Military Medical Corps: 1916–1966," *JAOA* 98 (1998): 279–86.

44. "Osteopathy's Epidemic Record," *OP* 36 (July 1919): 1; "Death Statistics Reveal Comparative Values of Osteopathic and Drug Treatments," *OP* 34 (December 1918): 1–2.

45. "Editorial: Figures Never Lie," *JAMA* 72 (1919): 731.

46. "The Profession's Policy," *JAOA* 19 (1920): 482–83; "Report of the Associated Colleges," *JAOA* 19 (1920) supplement: 6–7.

47. As of 1927 only eleven states gave DOs the same unlimited license privileges granted to MDs. See American Osteopathic Association, *Digest of State Laws Relating to Osteopathy* (Chicago, 1927).

48. E. S. Comstock, "A Professional Problem," *JAOA* 23 (1924): 524; idem, "Chicago College Curriculum," *JAOA* 24 (1925): 460; Asa Willard, *"Materia Medica* in the Colleges," *JAOA* 25 (1925): 279.

49. George Carpenter, "Between the Devil and the Deep Blue Sea, or Damned If We Do and Damned If We Don't," *Forum of Ost* 2 (December 1928): 3–4.

50. "Proceedings of the House of Delegates," *Forum of Ost* 3 (August 1929) supplement: 5.

51. "Correspondence: Materia Medica," *Forum of Ost* 2 (October 1928): 14–15; and *Forum of Ost* 2 (November 1928): 10–11; H. L. Knapp, "An Open Letter to Dr. John A. MacDonald," *Forum of Ost* 2 (December 1928): 3; Warren Davis, "People Want Osteopathy," *Forum of Ost* 2 (February 1929): 7.

52. "Proceedings of the House of Delegates," *Forum of Ost* 3 (August 1929) supplement: 6.

53. "Proceedings of the House of Delegates," *Forum of Ost* 4 (October 1930) supplement: 6–8.

CHAPTER 6. THE PUSH FOR HIGHER STANDARDS

1. See Paul Starr, *The Social Transformation of American Medicine* (New York: Basic Books, 1982) pp. 102–16.

2. John Field, "Medical Education in the United States: Late Nineteenth and Twentieth Centuries," in *The History of Medical Education*, ed. C. D. O'Malley (Berkeley: University of California Press, 1970), pp. 508–9. See also V. Johnson and H. G. Weiskotten, *A History of the Council on Medical Education and Hospitals of the American Medical Association* (Chicago: American Medical Association, 1960).

3. Abraham Flexner, *Medical Education in the United States and Canada: A Report to the Carnegie Foundation for the Advancement of Teaching* (Boston: Merrymount Press, 1910), pp. 62–67, 91–103.

4. Ibid., p. 151.

5. See Kenneth Ludmerer, *Learning to Heal: The Development of American Medical Education* (New York: Basic Books, 1985), pp. 173–90; William Rothstein, *American Medical Schools and the Practice of Medicine* (New York: Oxford University Press, 1987) pp. 142–49.

6. There were 95 MD-granting schools in the United States in 1915; 85 in 1920; and 71 in 1927. See Morris Fishbein, *A History of the American Medical Association, 1847 to 1947* (Philadelphia: W. B. Saunders, 1947), p. 898.

7. H. G. Weiskotten et al., *Medical Education in the United States, 1934 to 1939* (Chicago: American Medical Association, 1940), p. 15.

8. Ibid., pp. 67, 107.

9. Saul Jarcho, "Medical Education in the United States, 1910–1956," *Journal of the Mount Sinai School of Medicine* 26 (1957): 351; Field, "Medical Education," pp. 512–13.

10. Of the 5,611 MDs graduating in 1935, 5,491 (98 percent) immediately entered approved internships. See "Medical Education in the United States and Canada: Data for the Academic Year 1935–36," *JAMA* 107 (1936): 669.

11. Minutes of the December 28, 1908 meeting of the AMA Council on Medical Education, Archives of the American Medical Association, Chicago. See Norman Gevitz, "The Fate of Sectarian Medical Education" in *Beyond Flexner: Medical Education in the Twentieth Century*, ed. Barbara Barzansky and Norman Gevitz (Westport, Conn.: Greenwood Press, 1992), pp. 83–97.

12. Flexner, *Medical Education*, p. 164.

13. Ibid., pp. 164–66.

14. Ibid., p. 166.

15. "Carnegie Foundation Report," *JAOA* 10 (1911): 621–22.

16. "Report of the Committee on Education," *JAOA* 10 (1910): 35–38.

17. John Rogers, "Report of the Committee on College Instruction," *JAOA* 31 (1932): 509; Completed AOA Survey Questionnaires of Osteopathic Colleges for the academic year 1931–32, microfilmed, American Osteopathic Association Archives, Chicago.

18. George Laughlin, "Hindrances of Osteopathic Progress," *JAOA* 24 (1925): 519. Upon Andrew Still's death in 1917 his grandnephew George Still became president. A family struggle ensued, resulting in some members' leaving the American School of Osteopathy. In 1922 Laughlin, an orthopedic surgeon, opened a rival school in Kirksville, the Andrew Taylor Still College of Osteopathy and Surgery. Upon George Still's death that same year, negotiations were begun for the merger of the two institutions, which was effected in 1924. See E[amons] R. Booth, *History of Osteopathy and Twentieth-Century Medical Practice* (Cincinnati: Caxton Press, 1924), pp. 548–49; Georgia Warner Walter, *The First School of Osteopathic Medicine* (Kirksville, Mo.: Thomas Jefferson University Press, 1992) pp. 93–164.

19. Laughlin, "Hindrances to Osteopathic Progress," p. 519.

20. R. H. Singleton, "Report of Committee on American Osteopathic Foundation," *JAOA* 31 (1932): 511. See also John Rogers, "Report of the Bureau of Professional Education and Colleges," *JAOA* 36 (1937): 607.

21. See Asa Willard, "State Legal and Legislative Matters," *Forum of Ost* 4 (October 1930) supplement: 12.

22. This school is not to be confused with the Central College of Osteopathy, also of Kansas City.

23. Carol Benenson Perloff, *To Secure Merit: A Century of Philadelphia College of Osteopathic Medicine: 1899–1999* (Philadelphia: Philadelphia College of Osteopathic Medicine, 1999), pp. 22–27.

24. For example, see A. W. Bailey, "Osteopathic Education," *JAOA* 24 (1925): 355–58.

25. "Medical Education in the United States and Canada: Data for the Academic Year 1935–36," pp. 684–85.

26. Data derived from Completed AOA Survey Questionnaires of Osteopathic Colleges for the academic year 1935–36, microfilmed, American Osteopathic Association Archives, Chicago.

27. Data derived from college catalogs.

28. College of Osteopathic Physicians and Surgeons, *Annual Announcement* (1935), p. 71.

29. See John Wood, "Public Tax Supported Hospitals," *JAOA* 50 (1951): 141–44.

30. George Woodbury, "Unit Number Two of the Los Angeles County General Hospital: What It Is and How It Came About," *Western Osteopath* 23 (September 1928): 7–11.

31. Precisely how many osteopathic hospitals there were in this period is not known, since such institutions were under no obligation to identify themselves to the AOA. See Edgar Holden, "Report of the Bureau of Hospitals," *JAOA* 35 (1935): 46.

32. John Rogers, "Report of Bureau of Professional Education and Colleges," *JAOA* (1932): 508.

33. As derived from American Osteopathic Association, *Abstract of Laws Governing the Practice of Osteopathy* (Chicago, 1937), pp. 3–15.

34. Osteopathic data from "Report of the American Association of Osteopathic Examiners: 1952," microfilmed, American Osteopathic Association Archives, Chicago; MD data derived from "Medical Education in the United States," *JAMA* 90 (1928): 1203; *JAMA* 92 (1929): 1434–35; *JAMA* 94 (1930): 1312–13; *JAMA* 96 (1931): 1392–93; *JAMA* 98 (1932): 1460–61.

35. In 1907, 40.5 percent of all AOA-listed DOs were located in five states that had colleges: California, Illinois, Iowa, Missouri, and Pennsylvania. In 1940 this figure stood at 42.1 percent. Data derived from American Osteopathic Association, *Yearbook and Directory* (Chicago, 1940).

36. Norman Gevitz, "'A Coarse Sieve': Basic Science Boards and Medical Licensure in the United States," *J Hist Med* 43 (1988): 36–63; and Robert Derbyshire, *Medical Licensure and Discipline in the United States* (Baltimore: Johns Hopkins Press, 1969), pp. 118–33.

37. As quoted in Asa Willard, "Basic Science Boards," *Forum of Ost* 2 (November 1928): 2.

38. "State Board Statistics for 1930," *JAMA* 96 (1931): 1399.

39. Willard, "State Legal and Legislative Matters," p. 13.

40. As quoted in Willard, "Basic Science Boards," p. 3.

41. Frederick Etherington and S. Stanley Ryerson, "Preliminary Report to the Joint Advisory Committee Representing the College of Physicians and Surgeons of Ontario, the Ontario Medical Association, and the Universities in Ontario Engaged in the Teaching of Medicine on Osteopathic Colleges and Teaching in Kirksville, Philadelphia, Des Moines, and Chicago," March 1, 1934, microfilmed, American Osteopathic Association Archives, Chicago.

42. See "Report of the Council on Medical Education and Hospitals," *JAMA* 114 (1940): 1926.

43. See Ray Hulburt, "The Ontario Investigation of Osteopathy," *JAOA* 34 (1935): 466–71. Although the official report of Etherington and Ryerson was straightforward and factual, other statements by them tended to support the DOs' charges of bias. See Frederick Etherington, "Osteopathy and Licensure," *JAMA* 104 (1935): 1549–52.

44. L. E. Blauch, "Studies of the Chicago, Des Moines, Kansas City and Philadelphia Osteopathic Colleges," 1936, microfilmed, American Osteopathic Association Archives, Chicago. Blauch's Kirksville study is not included in the archival collection.

45. "Entrance Requirements, Enrollments, Next Steps," *JAOA* 39 (1939):

225–26. In 1943 the AOA began enforcing a requirement that the matriculants had to already have taken a minimum number of courses in English, biology, physics, and chemistry. See R. McFarlane Tilley, "Report of the Bureau of Professional Education and Colleges," *JAOA* 93 (1943): 81–83.

46. Asa Willard, "Where Our Students Come From," *JAOA* 46 (1947): 313.

47. Lawrence Mills, "Colleges Visited," *JAOA* 45 (1946): 422; *JAOA* 46 (1947): 591–92,

48. See Lawrence Mills, "Applications to Osteopathic Colleges," *JAOA* 51 (1952): 541–42.

49. Lawrence Mills, "Osteopathic Education," *JAOA* 50 (1951): 277–78.

50. Data derived from college catalogs.

51. Data derived from abstracted minutes of the sessions of the American Association of Osteopathic Colleges, 1945–60, microfilmed, American Osteopathic Association Archives, Chicago.

52. Data derived from college catalogs.

53. Data derived from Completed AOA Hospital Questionnaires for the year 1960, microfilmed, American Osteopathic Association Archives, Chicago.

54. Data derived from college catalogs.

55. "Osteopathic Progress Fund Reaches $962,535 as of June 15th," *JAOA* 43 (1944): 527.

56. Figures derived from "Recap of Annual Cash Received by the Osteopathic Progress Fund, 1946–1975," typescript, American Osteopathic Association Archives, Chicago.

57. "Cancer Teaching Grants to Osteopathic Colleges," *JAOA* 51 (1951): 126. "Six Colleges Report USPHS Grants," *Forum of Ost* 30 (1956): 292. "Hospital Survey and Construction Act," *JAOA* 46 (1946): 24–26; "Important Change in Hospital Construction Act Regulations," *JAOA* 46 (1947): 570.

58. See American Osteopathic Association, *Standardization of Osteopathic Hospitals, Including Codes, Hospital Regulations, Requirements for Teaching of Interns*, 2nd ed. (Chicago: AOA, 1938).

59. R. C. McCaughan, "Report of the Executive Secretary," *JAOA* 45 (1945): 23.

60. Floyd Peckham, "Report of the Bureau of Hospitals," *JAOA* 51 (1951): 74.

61. "AOA History: Dates, Events, and People," *JAOA* 77 (April 1978) supplement: 10.

62. R. McFarlane Tilley, "Report of the Advisory Board for Osteopathic Specialists," *JAOA* 39 (1939): 74–75.

63. See R. C. McCaughan, "Report of the Executive Secretary," *JAOA* 42 (1942): 46; R. McFarlane Tilley, "Report of the Bureau of Professional Education and Colleges," *JAOA* 46 (1946): 75; *JAOA* 47 (1947): 78–79; *JAOA* 49 (1949): 57; *JAOA* 50 (1950): 76. Also see Robert Thomas, "Report of the Bureau of Professional Education and Colleges," *JAOA* 53 (1953): 75. American Osteopathic Association, *Abstract of Laws and Regulations Governing the Practice of Osteopathy* (Chicago, 1960), p. 2.

1. Robert Thomas, "Report of the Council on Education," *JAOA* 50 (1950): 88.

2. G. W. Woodbury, "The Treasure of Distinctive Osteopathy," *JAOA* 39 (1940): 367.

3. E. A. Ward, "Pneumonia: Comparative Therapeutics," *JAOA* 49 (1950): 318–20; Floyd Peckham, "How to Obtain Better Cooperation between the Profession and the Hospital," *JAOA* 45 (1946): 199–200.

4. C. Robert Starks, "Our Greatest Challenge," *JAOA* 45 (1946): 537.

5. Floyd Peckham, "Report of the Bureau of Hospitals," *JAOA* 51 (1951): 74.

6. See Raymond Keesecker, "The Road Ahead for Osteopathy," *Forum of Ost* 29 (1955): 283; Woodbury, "Treasure of Distinctive Osteopathy," p. 367; Stanley Evans, "Future of Osteopathy," *Osteopathic Profession* 15 (March 1948): 12.

7. See J. McKee Arthur, "The Editor's Page," *Osteopathic Profession* 9 (1942): 28–29.

8. Henry Bunting, "Finding Ourselves in This Antitoxin Problem," *OP* 29 (January 1916): 2–3. In 1924 Bunting credited the late Michael A. Lane, a professor at Kirksville with writing this and other editorials on the value of antitoxin under Bunting's name. The identification of Lane as the author would likely have caused his removal from the faculty. Lane's claim to fame, which was made prior to his coming into connection with the osteopathic profession, was being the first to identify and name the alpha and beta cells of the Islets of Langerhans, a significant milestone on the way to a greater understanding of the cause and treatment of diabetes. Though an accomplished investigator, Lane's services as a researcher were rejected by the A. T. Still Research Institute, quite possibly because of his liberal views on serums and vaccines. See Henry Bunting, "Medical Recognition of Professor M. A. Lane's Contribution to Research," *OP* 46 (September 1924): 2–3. See also Robert Nichols, "Editorial: Reasonable Arguments," *Osteopathic Research Internist* 1 (December 1924): 163–66.

9. J. Stedman Denslow, "Guest Editorial: Ralph Waldo Gerard, Distinguished and Courageous Scientist," *JAOA* 73 (1974): 793–96.

10. J. Stedman Denslow and G. H. Clough, "Reflex Activity in the Spinal Extensors," *Journal of Neurophysiology* 4 (1941): 430–37; J. Stedman Denslow and C. C. Hassett, "The Central Excitatory State Associated with Postural Abnormalities," *Journal of Neurophysiology* 5 (1942): 393–402; J. Stedman Denslow and C. C. Hassett, "The Polyphasic Action Currents of the Motor Unit Complex," *American Journal of Physiology* 139 (1943): 652–59; J. Stedman Denslow, "Analysis of Variability of Spinal Reflex Thresholds," *Journal of Neurophysiology* 7 (1944): 207–15.

11. J. Stedman Denslow, Irwin Korr, and A. D. Krems, "Quantitative Studies of Chronic Facilitation in Human Motor Neuron Pools," *American Journal of Physiology* 105 (1947): 229–38. Korr continued his research career into the 1970s. See in particular I. M. Korr, P. N. Wilkinson, and F. W. Chornick,

"Axonal Delivery of Neuroplasmic Components to Muscle Cells," *Science* 155 (1967): 342–45.

12. See W. V. Cole, *An Introduction to Osteopathic Medicine* (Kansas City, Mo.: Kansas City College of Osteopathy and Surgery, 1961), pp. 64–65.

13. Louis Chandler, "Physiological Integration as a Basis for Recovery from Disease and Its Osteopathic Implication," *JAOA* 49 (1950): 305–15.

14. For a more detailed summary of pharmocotherapeutic advances during this era, see Ernst Baumler, *In Search of the Magic Bullet* (London: Thames and Hudson, 1965); L. Earle Arnow, *Health in a Bottle* (Philadelphia: J. B. Lippincott, 1970); and John Mann, *The Elusive Magic Bullet: The Search for the Perfect Drug* (New York: Oxford University Press, 1999).

15. Minutes of the AOA Board of Trustees, July 19–23, 1948, microfilmed, American Osteopathic Archives, Chicago, pp. 72–73.

16. In 1951 the AOA House of Delegates passed a resolution urging the various boards of specialty certification to "insist upon a well developed understanding of osteopathic principles and a demonstrated ability to apply those principles as a primary prerequisite for certification as an osteopathic specialist." See "Proceedings of the House of Delegates," *JAOA* 51 (1951): 25.

17. Lawrence Mills, "Adequacy of Undergraduate Osteopathic Teaching as Judged by Osteopathic Physicians Who Graduated from 1948 through 1953," microfilmed, American Osteopathic Association Archives, Chicago.

18. Margaret Barnes, "A Fortieth Anniversary Memoir," *The DO* 18 (January 1978): 25–29; Margaret Barnes, "History of the Academy of Applied Osteopathy," *The DO* 12 (June 1972): 113–33.

19. William Garner Sutherland, *The Cranial Bowl* (USA: Free Press, 1948; reprint of 1st ed.).

20. Ibid., p. 67.

21. For the current claims and supporting evidence as put forward by its advocates see Hollis King and Edna Lay, "Osteopathy in the Cranial Field," *Foundations for Osteopathic Medicine*, ed. Robert C. Ward (Philadelphia: Lippincott, Williams and Wilkins, 2002), pp. 985–1001.

22. Data culled from American Osteopathic Association, *Yearbook and Directory* (Chicago, 1960).

23. "The Man on the Street Gives His Ideas on Osteopathy," *Forum of Ost* 11 (1937): 35, 51–52.

24. George Hartman, "The Relative Social Prestige of Representative Medical Specialties," *Journal of Applied Psychology* 20 (1936): 659–63.

25. Alice Foley, "Osteopaths or Osteopathic Physicians," *JAOA* 25 (1926): 371; M. F. Hulett, "Osteopathic Physician and Surgeon," *JAOA* 25 (1926): 458; Cyrus Gaddis, "Away from Congested Centers," *JAOA* 26 (1926): 204.

26. "New Medical Dictionary Defines Osteopathy," *Forum of Ost* 10 (1936): 205; Ray Hulburt, "Definitions—Spinal Joints—Osteopathic Physicians," *Forum of Ost* 5 (1931): 194–95; George McCole, "Osteopathic Definitions," *Forum of Ost* 10 (1936): 151, 168.

27. "Proceedings of the House of Delegates," *Forum of Ost* 3 (August 1929): 6; "Osteopathic Physicians and Directories," *Forum of Ost* 1 (December 1927): 12; R. C. McCaughan, "Report of the Executive Secretary," *JAOA* 36 (1937): 594–96.

28. "Federal Emergency Sick Relief," *Forum of Ost* 7 (1933): 183–84; "FERA and CWA," *Forum of Ost* 7 (1934): 263; "Where Do We Go from Here?" *Forum of Ost* 12 (1938): 97, 114; B. F. Adams, "Report of the Committee on Compensation Insurance," *JAOA* 26 (1946): 93, 94; Robert Homan, "Report of the Committee on Compensation Insurance," *JAOA* 48 (1948): 69; John Wood, "The Audrain County Hospital Case," *JAOA* 50 (1951): 292–93; Don Cameron, "Can a Hospital Survive a DO Invasion?" *Medical Economics* 30 (July 1953): 99–105; "Hospital Survey and Construction Act," *JAOA* 46 (1946): 24–26; "Important Change in Hospital Construction Act Regulations," *JAOA* 46 (1947): 570.

29. Mark Sullivan, "If I Need Relaxation," *Reader's Digest* 34 (February 1939): 86–88.

30. See Jackson Harrison Pollack, *Dr. Sam: An American Tragedy* (Chicago: Henry Regnery, 1972); and Cynthia Cooper and Sam Reese Sheppard, *Mockery of Justice: The True Story of the Sheppard Murder Case* (Boston: Northeastern University Press, 1995).

31. Raymond Keesecker, "To the Student Wife," *Form of Ost* 29 (1955): 322–23. See also Peter K. New, "The Osteopathic Students: A Study in Dilemma," in *Patients, Physicians and Illness*, ed. E Gartly Jaco (Glencoe: Free Press, 1958), pp. 413–21.

32. O. W. Barnes, "Fifty Years Forecast of Osteopathy," *Forum of Ost* 3 (1929): 119; "Heresy or Science? Should We Award MD Degrees?" *Forum of Ost* 9 (1935): 29–31, 38.

33. R. McFarlane Tilley, "Report of the Bureau of Professional Education and Colleges," *JAOA* 41 (1941): 58, 59.

34. Donald Lewis, "DO and MD," *Forum of Ost* 1 (May 1927): 20; Abridged Proceedings, Mid-Year Meeting of the Executive Committee of the Board of Trustees, December 18–20, 1942, microfilmed, American Osteopathic Association Archives, Chicago, pp. 35, 132–33.

35. This failure to identify one's osteopathic affiliation extended to the hospitals. However, in 1947 the AOA house mandated that all such facilities include either the word *osteopathy* or the word *osteopathic* in their title or subtitle. See Floyd Peckham, "Report of the Bureau of Hospitals," *JAOA* 47 (1947): 92.

CHAPTER 8. THE CALIFORNIA MERGER

1. See Louisa Bartosh, "The History of Osteopathy in California," *Journal of the Osteopathic Physicians and Surgeons of California* 5 (April–May 1978): 30–33.

2. "Osteopathic Unit Makes Excellent Showing in Annual Report," *Western Osteopath* 24 (September 1929): 7–10; G. W. Woodbury, "The College and Unit #2," *Western Osteopath* 24 (May 1930): 11–14; Louis Chandler, "Progress and Problems at Unit #2," *Western Osteopath* 26 (March 1932): 7; L. C. Chandler,

"Fishbein vs. the Osteopathic Unit," *Western Osteopath* 27 (November 1932): 5–6; Dain Tasker Historical Manuscript, chapters 41–43, Special Collections, University of California-Irvine.

3. W. Ballentine Henley, "Comes the Dawn," *JAOA* 58 (1958): 141–47.

4. "Proceedings of the House of Delegates," *JAOA* 40 (1940): 33–34.

5. Dain Tasker Historical Manuscript, chapter 58.

6. For a discussion of the absorption of the homeopaths and eclectics, see William Rothstein, *American Physicians in the Nineteenth Century: From Sects to Science* (Baltimore: Johns Hopkins University Press, 1972), pp. 298–326.

7. Arnold Kisch and Arthur Viseltear, *Doctors of Medicine and Doctors of Osteopathy in California: Two Medical Professions Face the Problem of Providing Medical Care* (Arlington, Va.: Department of Health, Education and Welfare, Public Health Service, Division of Medical Care Administration, 1967), pp. 14–15.

8. Minutes of the California Osteopathic Association House of Delegates and Board of Trustees, March 18–19, 1944, microfilmed, American Osteopathic Association Archives, Chicago, pp. 5–6.

9. Ibid., p. 7; Kisch and Viseltear, *Doctors of Medicine*, p. 15.

10. Minutes of the California Osteopathic Association House of Delegates and Board of Trustees, March 18–19, 1944, p. 7; Minutes of the California Osteopathic Association, Report of the Fact-Finding Committee, March 3–4, 1945, microfilmed, American Osteopathic Association Archives, Chicago.

11. Metropolitan University, College of Medicine and Surgery, Graduate Division, *Annual Catalog* (Los Angeles, 1945); Metropolitan University File, microfilmed, American Osteopathic Association Archives, Chicago.

12. Minutes of the American Osteopathic Association Board of Trustees, July 1946, microfilmed, American Osteopathic Association Archives, Chicago, pp. 171–75.

13. A detailed file of meetings, reports and letters related to the discussions between representatives of the COA and the CMA can be found in the Forest Grunigen Papers, Special Collections, University of California-Irvine. See also Dolores Grunigen and Jay O'Connell, *A Strength Born of Giants: The Life and Times of Dr. Forest Grunigen* (Van Nuys, Calif.: Raven River Press, 2002), pp. 51–95.

14. Grunigen and O'Connell, *A Strength Born of Giants*, 51–95.

15. Minutes of the American Osteopathic Association Board of Trustees, July 11–15, 1949, microfilmed, American Osteopathic Association Archives, Chicago, pp. 201–2; Minutes of the California Osteopathic Association House of Delegates and Board of Trustees, April 27–28, 1949, microfilmed, American Osteopathic Association Archives, Chicago, p. 15.

16. Facts Relating to the Origins of the AOA-AMA Conference Committee Meetings, n.d., microfilmed, American Osteopathic Association Archives, Chicago, p. 4.

17. Kisch and Viseltear, *Doctors of Medicine*, pp. 15–16.

18. Ibid., pp. 5–6.

19. "Editorial: The AOA and AMA Conferences," *Forum of Ost* 27 (1953): 186–87.

20. Wire Recording Notes Taken March 8, 1952, by the American Osteopathic Association Conference Committee, microfilmed, American Osteopathic Association Archives, Chicago.

21. "Editorial: The AOA and AMA Conferences," p. 187.

22. Ibid.

23. "Address of the President, Dr. John W Cline," *JAMA* 149 (1952): 853–56; "Report of the Reference Committee on Miscellaneous Business," *JAMA* 149 (1952): 944; "Report of Officers," *JAMA* 150 (1952): 892; "Report of the Judicial Council," *JAMA* 150 (1952): 1706.

24. "Editorial: The AOA and the AMA Conferences," p. 188.

25. "Report of the Committee for the Study of Relations between Osteopathy and Medicine," *JAMA* 152 (1953): 734–39.

26. Report of the American Osteopathic Association Conference Committee to the A.O.A. House of Delegates, July 1954, microfilmed, American Osteopathic Association Archives, Chicago, p. 2.

27. Ibid., pp. 2–7.

28. Ibid., pp. 11–17.

29. Ibid., pp. 17–19.

30. Minutes of the American Osteopathic Association House of Delegates, July 1954, microfilmed, American Osteopathic Association Archives, Chicago, 281.

31. "Supplementary Report of the Board of Trustees," *JAMA* 156 (1954): 1600–1605; "The AOA and AMA Conferences: To Settle with Finality," *Forum of Ost* 28 (1954): 611–14.

32. Report of the Committee for the Study of Relations between Osteopathy and Medicine," *JAMA* 158 (1955): 736–42.

33. Ibid., p. 740.

34. Ibid., p. 741.

35. Ibid., pp. 41 42.

36. Minutes of the Meeting of the American Osteopathic Association Conference Committee, June 11, 1955, microfilmed, American Osteopathic Association Archives, Chicago, pp. 1–5.

37. "The AOA and AMA Conferences: Settled without Finality," *Forum of Ost* 29 (1955): 244–47.

38. True B. Eveleth to Floyd Peckham, November 21, 1957, microfilmed, American Osteopathic Association Archives, Chicago.

39. "Proceedings of the House of Delegates," *JAOA* 57 (1957): 68.

40. Ibid.

41. "Highlights of the Atlantic City Meeting," *JAMA* 168 (1958): 2150.

42. "Report of the Judicial Council," *JAMA* 171 (1959): 978–79.

43. "Highlights of the Atlantic City Meeting," *JAMA* 170 (1959): 1075; "MDs Can Teach DOs—If," *American Osteopathic Association News Bulletin* (hereafter *AOA News Bull*) (June 1959): 1–2.

44. "Remarks of George W. Northup, DO, to the House of Delegates, July 12, 1959," microfilmed, American Osteopathic Association Archives, Chicago, pp. 1–5.

45. "Text of Michigan Resolution," *AOA News Bull* 3 (August 1959): 1.

46. Forest Grunigen Papers; Kisch and Viseltear, *Doctors of Medicine*, pp. 23–24.

47. Ibid., pp. 25–27; "AOA Acts on Unity Talks," *AOA News Bull* 3 (August 1960): 1.

48. "Text of Michigan Resolution," p. 3.

49. Kisch and Viseltear, *Doctors of Medicine*, p. 28; "COA Charter Revoked," *AOA News Bull* 3 (November 1960): 1.

50. "AOA Charters New Group," *AOA News Bull* 4 (February 1961): 1.

51. Kisch and Viseltear, *Doctors of Medicine*, p. 31.

52. "A Report to the Membership," *JAOA* 60 (1961): 671–74.

53. Kisch and Viseltear, *Doctors of Medicine*, pp. 34–36.

54. Ibid., pp. 33–34.

55. "California Merger Program: Important Dates," microfilmed, American Osteopathic Association Archives, Chicago.

CHAPTER 9. REAFFIRMATION AND EXPANSION

1. "Osteopathy: Special Report of the Judicial Council to the AMA House of Delegates," *JAMA* 177 (1961): 775.

2. "Editorial: Osteopaths vs. Osteopathy," *JAMA* 177 (1961): 779.

3. "Osteopathy: Special Report of the Judicial Council," p. 775.

4. "Trustees Statement on AMA Policy," *AOA News Bull* 4 (July 1961): 3.

5. Ibid. Following this line of reasoning, economist Erwin Blackstone argued that the principal motivation behind this and other AMA policies towards the DOs was a desire to eliminate a viable competitor. See his "The AMA and the Osteopaths: A Study of the Power of Organized Medicine," *The Antitrust Bulletin* 22 (1977): 405–40. See also Howard Wolinsky and Tom Brune, *The Serpent on the Staff: The Unhealthy Politics of the American Medical Association* (New York: G. P. Putnam's Sons, 1994) pp. 121–43.

6. "Medical Societies Confer with Osteopaths," *AMA News* 8 (March 1, 1965): 2.

7. "Washington State MD-DO Plan Told," *AMA News* 6 (November 11, 1963): 16; "MD-DO Merger Efforts Continue," *AMA News* 7 (April 13, 1964): 16; "MD Degrees for Osteopaths Validated," *AMA News* 10 (January 9, 1967): 9; "Osteopaths' MD Degrees Denied," *AMA News* 11 (January 8, 1968): 1, 11.

8. "DOs Attend Town Hall Meeting to Discuss Current AOA Policies," *AOA News Bull* 5 (May 1962): 1; "AOA Will Hold Town Hall Session at MAOP&S Meeting," *AOA News Bull* 5 (October 1962): 1.

9. Galen Young, "Message from the President of the AOA," *JAOA* 59 (1960): 487–88.

10. Arnold Kisch and Arthur Viseltear, *Doctors of Medicine and Doctors of Osteopathy in California: Two Medical Professions Face the Problem of Providing Medical Care* (Arlington, Va.: Department of Health, Education, and Welfare, Public Health Service, Division of Medical Care Administration, 1967), p. 40; "When DOs Become MDs," *Medical Economics* 40 (December 2, 1963): 62.

11. See Nancy Kaye, "DOs Turned MDs: How Are They Faring?" *Medical Economics* 40 (November 4, 1963): 115–25.

12. Kisch and Viseltear, *Doctors of Medicine*, p. 42.

13. Ibid., p. 41.

14. "Profile of a Merger: Responses to Questionnaires, Analyses, Comments Conducted November 17–20, 1965, in California by the A.O.A. Public Relations Department," microfilmed, American Osteopathic Association Archives, Chicago.

15. Ibid., p. 6.

16. For an internal biography of the new school by an early MD dean, as well as his history of COP&S see Warren Bostick, *College of Medicine, University of California-Irvine* (Irvine: n.p., 1994).

17. "Sues to Stop Recognition of Little md by New York," *AOA News Bull* 9 (February 1966): 5. In New York, however, as a result of an administrative decision by the State Department of Education, DOs who held 1961 California College of Medicine diplomas could, if they chose, be listed on their license as "DO-MD," though they were examined and licensed on the basis of their osteopathic credentials. Although this practice was stopped by the courts in 1968, fifteen DOs won the right to retain this designation. "Fifteen New 'DO-MDs' Result from New York's Ruling," *AMA News* 7 (February 17, 1964): 16; "Certificate Listing 'MD-DO' Prompts New York Lawsuit," *AMA News* 9 (February 28, 1966): 9; "Osteopaths' MD Degrees Denied," *AMA News* 11 (January 8, 1968): 1, 11; "Osteopaths Can Display 'MD,'" *AMA News* 13 (February 2, 1970): 14.

18. See Kaye, "DOs Turned MDs: How Are They Faring?"

19. Jack Leahy, "How DOs Feel about AOA-AMA Relations," *OP* 38 (July 1972): 28.

20. "AHA Changes Listing Criteria," *AOA News Bull* 3 (October 1960): 1.

21. "Joint Commission Okays Mixed Staffs," *AOA News Bull* 3 (October 1960): 1.

22. "Changes in DO Policy Opposed," *AMA News* 8 (May 3, 1965): 7; "Medicolegal Decisions," *AMA News* 11 (November 25, 1968): 13; "Oppressive Actions in Maryland, Nebraska Seek to Deny Full Licensing of DOs," *AOA News Bull* 8 (May 1965): 3.

23. "DOs Qualify for Positions with U.S. Civil Service," *AOA News Bull* 6 (May 1963): 2; "Order Armed Forces to Commission DOs as Medical Officers," *AOA News Bull* 9 (June 1966): 1; "U.S. Recognizes AOA Hospital Accreditation for Use in Medicare," *AOA News Bull* 9 (November 1966): 1, 8. See also Norman Gevitz, "The Sword and the Scalpel: The Osteopathic War to Enter the Military Medical Corps," *JAOA* 98 (1998): 279–86.

24. "DO Education Changes Urged," *AMA News* 10 (July 3, 1967): 1, 8. See also "House of Delegates Meets," *JAMA* 201 (1967): 38.

25. "House of Delegates Rebuffs 'Academic Piracy' of AMA," *AOA News Bull* 10 (August 1967): 1, 8.

26. See Carl Waterbury, "DO-MD: Some Guidelines," *OP* 35 (November 1969): 17–19. J. Dudley Chapman, "The Other Side of the Des Moines College Crisis," *OP* 36 (July 1970): 17–23; "Dr. Waterbury Speaks Out on COMS," *OP* 37 (May 1971): 21–23.

27. "AMA Offers Means for DO Membership," *AMA News* 11 (December 16, 1968): 1, 8.

28. "DOs Can Now Join the AMA," *AMA News* 12 (July 28, 1969): 6; "Eligibility of Osteopaths for County and State Medical Society Membership," *JAMA* 210 (1969): 1512.

29. "House of Delegates Takes Action to Resolve Conflict of Interest," *AOA News Bull* 11 (August 1968): 1; "AOA House of Delegates Reaffirms Separate Status for DOs," *American Osteopathic Association News Review* 12 (September 1969): 1–2.

30. Edward Crowell, "AOA House and Board Meet in Denver: Reaffirm Membership Policy, Approve Health Insurance Statement," *The DO* 12 (October 1971): 45–46, 56–58; idem, "AOA Board and House Meet: Take Significant Actions," *The DO* 14 (October 1973): 75–76.

31. Data furnished by the American Medical Association, Department of Membership, December 1978.

32. "AOA House of Delegates Reaffirms Separate Status," pp. 1–2.

33. Carolyn Cranford, "AOA-AAOC Meet in Chicago," *The DO* 10 (February 1970): 51.

34. Barbara Peterson, "AOA House of Delegates Moves to Serve the Public Health," *The DO* 11 (September 1970): 53.

35. "Editorials," *JAOA* 70 (1970): 104.

36. *Final Report of the Osteopathic Medical Manpower Information Project* (Washington, D.C.: American Association of Colleges of Osteopathic Medicine, 1977) p. 40; "Medical Education in the United States," *JAMA* 218 (1971): 1247–48; 222 (1972): 1011–12; 226 (1973): 945; 238 (1977): 2790–92; 240 (1978): 2842–44. In the year 1972 to 1973 there were 415 DOs in AOA-accredited hospital residencies. In the year 1976 to 1977 the figure was 531. See *Final Report of the Osteopathic Medical Manpower Information Project*, p. 41.

37. "Osteopathy Ruling Voided," *American Medical News* 16 (October 15, 1973): 1, 3.

38. Rosemary Stevens, *American Medicine and the Public Interest* (New Haven: Yale University Press, 1971), pp. 362–67; William Rothstein, *American Medical Schools and the Practice of Medicine* (New York: Oxford University Press, 1987), pp. 283–88; Kenneth Ludmerer, *Time to Heal: American Medical Education from the Turn of the Century to the Era of Managed Care* (New York: Basic Books, 1999), pp. 210–15.

39. Data culled from U.S. Department of Health, Education, and Welfare, Public Health Service, Health Resources Administration, *Health Professions Schools: Selected BHM Support Data F.Y. 1965–1976* (Washington, D.C.: U.S. Department of Health, Education, and Welfare, 1977), p. 120.

40. Institute of Medicine, *Costs of Education in the Health Professions* (Bethesda, Md.: National Academy of Sciences, U.S. Department of Health, Education, and Welfare, Public Health Service, Health Resources Administration, Bureau of Health Resources Development, 1974), pp. xiv, xviii.

41. Data culled from catalogs of the colleges for the respective years.

42. *Final Report of the Osteopathic Medical Manpower Information Project*, pp. 47–54.

43. "Michigan Plans New College," *AOA News Bull* 6 (May 1963): 1; "Michigan College Site Moved from Lansing to Pontiac," *AOA News Bull* 7 (December 1964): 1.

44. "Michigan College Moves Two Steps," *AOA News Bull* 8 (August 1965): 3; "Michigan Study Cites Role of DOs," *AOA News Bull* 4 (August 1961): 3; "Plan for Health Care Drafted in Michigan," *AOA News Bull* 5 (September 1962): 4.

45. "Michigan College Moves Two Steps," p. 3.

46. "Michigan DOs Vote 87 Percent against Merger," *AOA News Bull* 9 (April 1969): 1–2.

47. "Michigan to Establish Osteopathic School," *AOA News Review* 12 (August 1969): 1–2.

48. Ibid., p. 2; see also John Walsh, "Medicine at Michigan State," *Science* 177 (1972): 1085–87; *Science* 178 (1972): 36–39, 288–91, 377–80.

49. "TCOM Becomes State-Supported Medical School as Governor Briscoe Signs S.B. 216," *Texas Osteopathic Physicians Journal* 32 (July 1975): 12–13; "Basic Science Education Agreement Signed with State University," *The DO* 12 (April 1972): 171; "TCOM Receives Government Funding in Excess of $800,000," *The DO* 13 (October 1972): 201–2; "State Grants $3.4 Million Appropriation for TCOM," *The DO* 13 (August 1973): 202.

50. Robert Osborne, *A History of the Oklahoma State University College of Osteopathic Medicine* (Stillwater, Okla.: Oklahoma State University, 1998), pp. 29–123.

51. Carol Thiessen, "Greenbriar: The Little College that Could," *The DO* 15 (March 1975): 86–92; Penny Ellis and Alayne Steiger, *The DOs: Medicine in the Mountains* (Charleston: West Virginia Society of Osteopathic Medicine, 1986) pp. 100–108.

52. Carol Thiessen, "And Now There Are Ten," *The DO* 17 (September 1976): 85.

53. Jeff Kressman, "New Colleges Open in New Jersey, New York," *The DO* 18 (March 1978): 33–37.

54. Edward Crowell, "Accelerating Educational Growth Dominates Board Session," *The DO* 19 (October 1978): 61.

55. Carol Thiessen, "California Supreme Court Reopens the Golden State to Licensure," *The DO* 18 (April 1978): 46.

56. Allen Singer, *2000 Annual Statistical Report* (Bethesda, Md.: American Association of Colleges of Osteopathic Medicine, 2001) p. 13; U.S. Department of Health, Education, and Welfare, Public Health Service, Health Resources Administration, Office of Graduate Medical Education, *Interim Report of the Graduate Medical Education National Advisory Committee to the Secretary* (Hyattsville, Md.: Department of Health, Education, and Welfare, 1979), p. 161.

CHAPTER 10. IN A SEA OF CHANGE

1. General works on the origins of and early developments in Medicare and Medicaid include Rashi Fein, *Medicare, Medical Costs: The Search for a Health Insurance Policy* (Cambridge: Harvard University Press, 1986); Theodore Marmor,

The Politics of Medicare (London: Routledge and Kegan Paul, 1970); Sheri David, *With Dignity: The Search for Medicare and Medicaid* (Westport, Conn.: Greenwood Press, 1985); Robert Stevens and Rosemary Stevens *Welfare in America: A Case Study of Medicaid* (New York: Free Press, 1974).

2. For the method of calculating Medicare reimbursement of the costs of osteopathic internships and residencies see James Carl and Ronald Knaus, "A Primer on Graduate Medical Education Financing," *JAOA* 93 (1993): 1055–59.

3. See Odin Anderson, *Blue Cross Since 1929: Accountability and the Public Trust* (Cambridge, Mass.: Ballinger, 1975); Sylvia Law, *Blue Cross: What Went Wrong?* 2nd ed. (New Haven: Yale University Press, 1976); Robert Cunningham III and Robert Cunningham Jr., *The Blues: A History of the Blue Cross and Blue Shield System* (DeKalb: Northern Illinois University Press, 1997).

4. Alain Enthoven "A New Proposal to Reform the Tax Treatment of Health Insurance," *Health Affairs* 3(1984): 21–39.

5. Paul Starr, *The Social Transformation of American Medicine* (New York: Basic Books, 1982), p. 394.

6. Harry Sultz and Kristina Young, *Health Care USA: Understanding Its Organization and Delivery*, 3rd ed. (Gaithersburg, Md.: Aspen Publications, 2001), p. 230.

7. Rosemary Stevens, *In Sickness and in Wealth: American Hospitals in the Twentieth Century* (New York: Basic Books, 1989) pp. 307–9.

8. In 1970 the AOA House of Delegates unanimously supported the concept of comprehensive national health insurance. See *The DO* 11 (September 1970): 47–49. However, by 1979, experience with federal reimbursement policies and mandates was instrumental in the AOA's rescinding of its support. "National Health Policy Statement," *The DO* 20 (October 1979): 31–32.

9. Edward Crowell, "AOA Board and House Meet," *The DO* 14 (October 1973): 73; Wallace Pearson, "Council on Federal Health Programs," *The DO* 15 (October 1974): 163–64; Frank McDevitt, "Study Committee on HMOs, PSROs and Physicians' Unions," *The DO* 16 (October 1975): 117–18.

10. Donald Siehl, "Medical Care Criteria," *The DO* 17 (November 1976): 67–69.

11. John Perrin, "Washington Office," *The DO* 19 (November 1978): 131.

12. George Luibel, "Department of Public Affairs," *The DO* 15 (October 1974): 151–52; Edward Crowell, "Osteopathic Education Dominates Cincinnati Meeting," *The DO* 19 (October 1978): 64; "Plans Turned Down; DOMC Tries Again," *The DO* 19 (May 1978): 75, 77; "Community Champions Hospital Against Planning Agency," *The DO* 22 (June 1981): 85–87.

13. John Perrin, "Washington Office," *The DO* 21 (October 1980): 117.

14. John Perrin, "Washington Office," *The DO* 23 (October 1982): 150; Frank McDevitt, "Department of Government Affairs," *The DO* 24 (October 1983): 162; "New Osteopathic Hospital Opens in Florida," *The DO* 32 (February 1991): 106.

15. Sultz and Young, *Health Care USA*, p. 216.

16. Marcelino Oliva, "Bureau of Public Education on Health," *The DO* 25 (Oc-

tober 1984): 154; Elmer Baum, "Council on Federal Health Programs," *The DO* 25 (October 1984): 156; Rich Wolter, "Changes in PROs Designed to Better Relations with Physicians," *The DO* 32 (September 1991): 81–87.

17. Kenneth Thorpe and James Knickman, "Financing for Health Care," in *Health Care Delivery in the United States*, ed. Anthony Kovner and Steven Jonas (New York: Springer Publishing, 1999), pp. 54–56.

18. Sultz and Young, *Health Care USA*, pp. 110–35.

19. Lawrence Brown, *Politics and Health Care Organization: HMOs as Federal Policy* (Washington, D.C.: Brookings Institution, 1983); John B. McKinley, ed. *Health Maintenance Organizations* (Cambridge: MIT Press, 1981).

20. James Robinson, *The Corporate Practice of Medicine* (Berkeley: University of California Press, 1999).

21. Janet Horan, "National Study of the Impact of Managed Care on Osteopathic Physicians," *JAOA* 100 (2000): 218–27. See also Barbara Ross-Lee and Michael Weiser, "Managed Care: An Opportunity for Osteopathic Physicians," *JAOA* 94 (1994): 149–56; Daniel Bade, "AOA Mounts Response to Managed Care Industry," *The DO* 35 (February 1994): 57–60.

22. Michael Fitzgerald, "AOA Evaluates Values Assigned to OMT Codes," *The DO* 32 (April 1991): 76, 79. Daniel Bade, "Reimbursement for OMT Improves Under New MFS," *The DO* 33 (September 1992): 91–92.

23. Jennifer Cook and John Sprovieri, "Dr. Stowers Becomes First DO to Serve on PPRC," *The DO* 36 (May 1995): 29–30. In 2000 this DO became the first osteopathic physician to be appointed to MedPAC, the successor of the PPRC. See Heidi Ann Ecker, "Dr. Stowers Joins MedPAC as its First DO Member," *The DO* 41 (June 2000): 63–65.

24. Stacy Bohlen, "Battle to Correct OBRA Omission Continues," *The DO* 35 (December 1994): 28–40; idem, "Congress Corrects OBRA '90 Oversight," *The DO* 37 (November 1996): 48–53.

25. R. W. Hubbard, "Bureau of Public Education on Health," *The DO* 31 (October 1990): 31; Elizabeth Beckwith, "Department of Government Relations," *The DO* 34 (October 1993): 110–11; Anthony Minissale, "Bureau of State Government Affairs," *The DO* 36 (October 1995): 96.

26. In addition, although HCFA issued a policy declaring that DOs could be reimbursed for both evaluation and management (EM) and OMT performed during the same visit, little implementation occurred at the Medicare carrier level. See "AOA Moves to Address OMT Reimbursement Problems," *The DO* 34 (April 1993): 33; John Sprovieri, "AOA Prompts HCFA to Issue Statement on OMT Codes," *The DO* 35 (November 1994): 80–82; Jennifer Berger, "Get What You Deserve: Perseverance, Saavy Key to Obtaining Reimbursement for OMT," *The DO* 39 (May 1998): 62–70; Amy Bennett, "Cacophony of Coding Can Be Cleared Up," *The DO* 42 (January 2001): 64–66.

27. "AOHA Releases Cost Data," *The DO* 18 (August 1978): 75; "Hospitals Growing—and So Are Costs," *The DO* 21 (March 1981): 121.

28. George Northup, "The Osteopathic Hospital," *JAOA* 80 (1980): 14–15,

96–97, 172–73, 244, 322; Bruce Krider, "Integrated Hospital Staffs: Will [They] Lead to DO-MD Amalgamation," *OP* 45 (August 1978): 43–45; Gilbert Toffol, "Saving DO Hospitals," *The DO* 30 (January 1989): 19.

29. "Osteopathic Organizations Seek Answers to Utilization Issue," *The DO* 25 (October 1984): 97; "DOs and DO Hospitals: A Loyalty Issue," *The DO* 27 (January 1986): 131–32.

30. "OP News Notes," *OP* 45 (August 1978): 6; "Osteopathic Hospitals Expand," *The DO* 22 (August 1982): 160–61.

31. Anne-Marie Roussel, "DRGs and Teaching Hospitals—An Uncertain Alliance," *The DO* 25 (November 1984): 105–6; "Hospital Sues Blue Cross for False Advertising," *The DO* 26 (February 1985): 13; John Sprovieri, "Hospitals Challenge Tough Economic Climate," *The DO* 29 (March 1988): 118–20; John Sprovieri, "Quality, Not DRGs, Should Guide Hospital Payment," *The DO* 30 (February 1989): 123–24.

32. Roger Zumwalt, "DRGs—Survival Strategies for Small Osteopathic Hospitals," *The DO* 25 (November 1984): 111–12; "Merge or Submerge?" *The DO* 26 (November 1985): 145–48; "Metropolitan Hospital Sells Its Divisions," *The DO* 31 (June 1990): 11; Richard Sims, "Osteopathic Hospitals Have Critical Choices to Make in the 1990s," *The DO* 32 (April 1991): 16–29; "Financial Woes Spur Detroit Osteopathic Hospital to Merge with Detroit Riverside," *The DO* 34 (February 1993): 123–24; "Columbia/HCA Acquires Chicago Osteopathic Hospitals," *The DO* 37 (April 1996): 66; Lydia Hodges, "Medical Mecca Remembered," *The DO* 40 (June 1999): 34–37; Carol Benenson Perloff, *To Secure Merit: A Century of Philadelphia College of Osteopathic Medicine, 1899–1999* (Philadelphia: Philadelphia College of Osteopathic Medicine, 1999) pp. 64–67.

33. Data derived from *AOA Yearbook and Directory of Osteopathic Physicians 1974* (Chicago: American Osteopathic Association, 1974), pp. 45–58; *AOA Yearbook and Directory of Osteopathic Physicians 1988–1989* (Chicago: American Osteopathic Association, 1989), pp. 595–604; *AOA Yearbook and Directory of Osteopathic Physicians 1999* (Chicago: American Osteopathic Association, 1999), pp. 730–35.

34. Mike Fitzgerald, "Specialty Hospitals: For Some an Alternative to Closing," *The DO* 29 (March 1988): 121–26; Ruth Mack, "Confronting Competition with Innovative Services," *The DO* 29 (March 1988): 127–29.

35. "Foundations Offer Future to Failing Hospitals," *The DO* 30 (May 1989): 116–21; David Rianda, "National Association of Osteopathic Foundations," *The DO* 32 (October 1991): 131–32; Carol Williams, "Foundations Play Critical Role in Promoting Profession," *The DO* 34 (November 1993): 82–86.

36. John Sprovieri, "Osteopathic Hospitals Strive to Maintain Identity in Changing Environment" *The DO* 36 (June 1995): 58–59; George Reuther, "Passing Muster: HFAP's AOA Program Ensures Quality of Healthcare Facilities," *The DO* 40 (June 1999): 20–26; "AOA First to Earn Authority to Accredit Critical Access Hospitals," *The DO* 43 (February 2002): 73.

37. *AOA Yearbook and Directory 1974*, pp. 45–58; *AOA Yearbook and Directory 1999*, pp. 730–35. Unfortunately these totals for recorded beds include hospitals that appear to be counting "licensed" beds as well as those institutions who limit

their bed totals to actual beds currently in service. In 2002, the American Osteopathic Health Care Association (AOHA), which represented member osteopathic hospitals in Washington for decades, ceased operations and its functions were taken over by the American Osteopathic Association.

38. Leonard Fenninger and Rose Tracy, "Graduate Medical Education," in *Medical Education in the United States, 1973–74*, ed. Anne Crowley (Chicago: American Medical Association, 1975), p. 35.

39. "Hospital Requirements for Intern Training and the Intern Registration Program," *JAOA* 76 (April 1977) supplement: 119–25.

40. Edward Crowell, "AOA Board and House Approve Special Assessment," *The DO* 18 (October 1975): 55–61; idem, "Board, House Respond to Small States' Concerns," *The DO* 18 (October 1977): 37–39.

41. "Task Force on Graduate Osteopathic Medical Education," *The DO* 21 (April 1981) supplement: 72–74.

42. Frank Campion, *The AMA and U.S. Health Policy* (Chicago: Chicago Review Press, 1984), pp. 447–52.

43. Mark Cummings, "Challenge to Osteopathic Education: Returning to its Primary Care Roots," *JAMA* 268 (1992): 1139–40; idem, "Combined Allopathic and Osteopathic GME Programs: A Good Thing, But Will They Continue?" *Academic Medicine* 74 (1999): 948–50.

44. Mark Cummings and Jan Wachtler, "Trends in Postdoctoral Training," *JAOA* 85 (1985): 722; Carolyn Swallow, Verna Bronersky, and Pamela Falbo, "Osteopathic Graduate Medical Education," *JAOA* 97 (1997): 647.

45. Michael Opipari, "Specialty-Oriented Internships," *JAOA* 94 (1994): 509–11; Helen Baker and Janice Wachtler, "Osteopathic Postdoctoral Education," *JAOA* 90 (1990): 1010–13; Swallow, Bronersky, and Falbo, "Osteopathic Graduate Medical Education," p. 647.

46. "Report of the Task Force to Explore Alternate Approval Mechanisms for Postdoctoral Training," *The DO* 29 (September 1988): 97–100; Swallow, Bronersky, and Falbo, "Osteopathic Graduate Medical Education," p. 657; Cummings and Wachtler, "Trends in Postdoctoral Training," p. 723.

47. In 1985–86, 12 percent of graduating DOs entered programs accredited by ACGME. In 1995–96, it was 23 percent. Unfortunately, it is not known how many of these ACGME programs were also accredited by the AOA. Baker and Wachtler, "Osteopathic Postdoctoral Education," pp. 1010–13; Carolyn Swallow, "Osteopathic Graduate Medical Education," *JAOA* 100 (2000): 682.

48. Swallow, Bronersky, and Falbo, "Osteopathic Graduate Medical Education," p. 657. See Andrew Pecora, "Factors Influencing Osteopathic Physicians' Decisions to Enroll in Allopathic Residency Programs," *JAOA* 90 (1990): 527–33. In 1988 the AOA changed its rules governing approval of nonfederal ACGME residency training. Until then, the AOA would not approve an ACGME residency taken by a DO after an AOA approved internship when there were available open positions in the same specialty within osteopathic programs. Graduates needed first to apply to currently active and funded osteopathic residencies. After 1988, DOs completing an AOA rotating internship could directly apply to the AOA

Committee on Postdoctoral Training for approval to take an ACGME residency. See "More Grads May Prompt Revised Residency Rules," *The DO* 29 (August 1988): 114–15.

49. Swallow, Bronersky, and Falbo, "Osteopathic Graduate Medical Education," pp. 646–47.

50. Michael Fitzgerald, "AOA Adopts Alternate Pathway for Board Certification," *The DO* 40 (June 1999): 18–19; "AOA Board Approves Alternatives for Obtaining AOA Approval of Internships," *The DO* 43 (March 2002): 8.

51. Christopher Meyer, "The Osteopathic Medical Game: New Strategies for Winning," *JAOA* 94 (1994): 715–31.

52. Christopher Meyer, "COGMET Offers Model for Osteopathic Consortia," *The DO* 32 (December 1991): 71–75.

53. Christopher Meyer, Ronald Portanova, Cheryl Riley, et al. "The Anatomy of an OPTI: Part 2. The CORE system," *JAOA* 97 (1997): 686–91.

54. Michael Opipari, "Osteopathic Postdoctoral Training Institution: The Osteopathic 'Road Map' to Graduate Medical Education Viability," *JAOA* 95 (1995): 666–67; Christopher Meyer, Mary Patt Mann, Cheryl Riley, et al. "Anatomy of an OPTI: Part 1. Form, Function, and Relationships," *JAOA* 97 (1997): 599–603; Bruce Bates, "OPTIs Pose Problems for Small States, Small Hospitals," *The DO* 36 (December 1995): 57–60; Dennis Agostino, "New College Limits Difficulties in Creating an OPTI," *The DO* 39 (December 1998): 63–66.

55. John Crosby, "Who Moved My Cheese? Optimizing OPTIs in the Post–BBA '97 Era," *The DO* 42 (August 2001): 11–13; Jed Magen, "Current Threats to Osteopathic Graduate Medical Education," *JAOA* 102 (2002): 156–60.

56. Oliver Hayes, "Dual Approval of a Residency Program: Ten Years Experience and Implications for Postdoctoral Training," *JAOA* 98 (1998): 647–52; Jill Svoboda, "As DO Graduates Multiply, AOA Pushes for New GME Sites," *The DO* 42 (February 2001): 58–61.

CHAPTER 11. THE CHALLENGE OF DISTINCTIVENESS

1. John Crosby, "'News' for 2000: New Hope, New Projects, New Opportunities," *The DO* 41 (January 2000): 19–20; Howard Levine, "Our Last Frontier," *The DO* 42 (September 2001): 56; "AOA Fact Sheet," *The DO* 41 (August 2000): 26–29; Donald Cherry and David Woodwell, "National Ambulatory Medical Care Survey: 2000 Summary," Advance Data, June 5, 2002, No. 328. (Washington, D.C.: National Center for Health Statistics), p. 3.

2. Jacquie Goetz, "Emerging States Fight to Make Name for Themselves," *The DO* 41 (November 2000): 43–47; "AOA Fact Sheet," pp. 28–29.

3. Allen M. Singer, *2001 Annual Report on Osteopathic Medical Education* (Chevy Chase, Md.: American Association of Colleges of Osteopathic Medicine, 2002), p. 22. "Osteopathic Medical College Established in Arizona," *The DO* 37 (April 1996): 16; "New Osteopathic School Opens," *The DO* 34 (July 1993): 9; James Castro, "A Medical School for the Mountains: Training Doctors for Rural Care," *Appalachia* 34 (September–December 2001): 24–29.

4. For the profession's response to the GMENAC report, which was the first of many studies calling for a cap on new and expanded medical schools, see George Northup, "Too Many or Too Few," *JAOA* 80 (1981): 387–88. For the location of DO graduates see Gary Gugelchuk and Judy Cody, "Physicians in Service to the Underserved: An Analysis of the Practice Locations of Alumni of Western University of Health Sciences College of Osteopathic Medicine of the Pacific, 1982–1995," *Academic Medicine* 74 (1999): 557–59; Allan Roberts et al., "An Approach to Training and Retaining Primary Care Physicians in Rural Appalachia," *Academic Medicine* 68 (1993): 122–25; R. Tennyson Williams, "Twenty Year Trends in the Ohio Generalist Physician Workforce," *Journal of Family Practice* 47 (1998): 434–39. See also E. S. Salsberg and G. J. Forte, "Trends in the Physician Workforce, 1980–2000," *Health Affairs* 21 (2002): 165–73.

5. "AOA Fact Sheet," p. 28; "The Third Annual Primary Care Scorecard," *The New Physician* 47 (April 1998): 13–15. Douglas Wood and Barbara Ross-Lee, "The Medical School Curriculum for the 21st Century," *JAOA* 98 (1998): 102–11.

6. For the 1999–2000 academic year there were an average of 69 full-time faculty members at DO schools compared to an average of 791 full-time faculty at MD schools. Unlike MD colleges, osteopathic institutions are heavily dependent upon part-time and voluntary clinical faculty to deliver their curricula. See Singer, *2001 Annual Report on Osteopathic Medical Education*, p. 50; and Le'Etta Robinson, *AAMC Data Book* (Washington, D.C.: American Association of Medical Colleges, 2000), p. 33.

7. John Lynch, "Clinical Research and the Osteopathic Profession," unpublished paper presented at the AOA Council on Federal Health Programs, September 2000, Washington, D.C. See also Norman Gevitz, "Researched and Demonstrated: Inquiry and Infrastructure at Osteopathic Institutions," *JAOA* 101 (2001): 174–78. In fiscal year 2000, data on the distribution of expenditures by function in osteopathic schools reveals that publicly financed osteopathic colleges devoted 7.4 percent of their expenditures to research compared to 1.4 percent by the private schools. See Singer, *2001 Annual Report*, p. 58.

8. Singer, *2001 Annual Report*, pp. 54–57. For an examination of recent challenges to allopathic medical education as a consequence of managed care see Kenneth Ludmerer, *Time to Heal* (New York: Oxford University Press, 1999).

9. Unpublished data furnished by the American Association of Colleges of Osteopathic Medicine, Chevy Chase, Md.

10. See for example "Chicago Osteopathic Health Systems Creates University," *The DO* 34 (June 1993): 30; "Nova Southeastern University Breaks Ground for Health Professions Complex," *The DO* 36 (March 1995): 78; "KCOM Builds Phoenix Training Facility," *The DO* 36 (January 1995): 10; "University in Des Moines Adopts New Name," *The DO* 40 (November 1999): 62.

11. The most serious threats of closure occurred in the freestanding publicly funded osteopathic schools in Oklahoma and West Virginia. See Robert E. Osborne, *A History of the Oklahoma State University College of Osteopathic Medicine* (Stillwater: Oklahoma State University, 1998), pp. 165–200; Penny Ellis and

Alayne Steiger, *The DOs: Osteopathic Medicine in the Mountains* (Charleston: West Virginia Society of Osteopathic Medicine, 1986), pp. 131–34; "Dr. Oliva Urges West Virginia to Continue School's State Funding," *The DO* 30 (April 1989): 13; "West Virginia Votes to Continue Sponsoring Osteopathic School," *The DO* 30 (May 1989): 13; "West Virginia School of Osteopathic Medicine Survives Merger Threat," *The DO* 32 (October 1991): 122.

12. Singer, *2001 Annual Report*, pp. 22–28. See also Douglas Eckberg, "The Dilemma of Osteopathic Physicians and the Rationalization of Medical Practice," *Social Science and Medicine* 25 (1987): 1111–20; Vladimir Shlapentokh, Neil O'Donnell, and Mary Beth Grey, "Osteopathic Interns' Attitudes Toward Their Education and Training," *JAOA* 91 (1991): 786–802; Margaret Aguwa and Daniel Liechty, "Professional Identification and Affiliation of the 1992 Graduate Class of the Colleges of Osteopathic Medicine," *JAOA* 99 (1999): 408–20.

13. Larry Besaw, "MDs and DOs: Allopaths and Osteopaths Have Learned to Coexist," *Texas Medicine* 93 (April 1997): 34–39; Lydia Hodges, "Symposium on Osteopathic Medicine Featured at Allopathic Convention," *The DO* 40 (May 1999): 31–32; Coimbra Sirica, ed., *Osteopathic Medicine: Past, Present, and Future* (New York: Josiah Macy Foundation, 1996); and idem, *Current Challenges to MDs and DOs* (New York: Josiah Macy Foundation, 1996).

14. For example see Brian Donadio, "Department of Government Relations," *The DO* 26 (October 1985): 178; Eugene Oliveri, "Bureau of State Government Affairs," *The DO* 34 (October 1993): 136–37.

15. Jennifer Berger, "State Examiners Discuss FSMB's Review of COMLEX-USA," *The DO* 40 (1999): 31–33; Gerald Osborn, "The Comprehensive Osteopathic Medical Licensing Examination, COMLEX-USA: A New Paradigm for Testing and Evaluation," *JAOA* 100 (2000): 1050–51; John Graneto, "Testing Osteopathic Medical School Graduates for Licensure: Is COMLEX-USA the Most Appropriate Examination?" *JAOA* 101 (2001): 26–32; Lydia Hodges, "While Seeking FSMB's Endorsement, COMLEX-USA Proves Its Validity," *The DO* 43 (May 2002): 52–54; Jill Svoboda, "Louisiana Governor Signs Legal Parity of DOs, MDs into Law," *The DO* 42 (July 2001): 41–42.

16. See Norman Gevitz, "The AMA and the Chiropractors: Reflections on the History of the Consultation Clause," *Perspectives in Biology and Medicine* 32 (1989): 281–99.

17. For example see Jeffrey Bouley, "Osteopathic Medicine's Distinction is EPPRC Bone of Contention," *The DO* 41 (October 2000): 49–51. Occasionally, DO leaders who strongly identify with the osteopathic profession declare that there is a strong likelihood that allopathic and osteopathic medicine will eventually merge into one profession. However, these expressions, made to MD audiences, are often diplomatic and perfunctory.

18. Data provided by the membership departments of the American Osteopathic Association and the American Medical Association.

19. See Norman Gevitz, "'Parallel and Distinctive': The Philosophical Pathway for Reform in Osteopathic Medical Education," *JAOA* 94 (1994): 328–32;

Wayne Guglielmo, "Are DOs Losing Their Unique Identity?" *Medical Economics* 75 (April 27, 1998): 201–14.

20. Special Committee on Osteopathic Principles and Osteopathic Technic, Kirksville College of Osteopathic Medicine, "An Interpretation of the Osteopathic Concept: Tentative Formulation of a Teaching Guide for Faculty, House Staff and Student Body," *J Ost* 60 (October 1953): 1–10; Sarah Sprafka, Robert Ward, and David Neff, "What Characterizes an Osteopathic Principle? Selected Responses to an Open Question," *JAOA* 81 (1981): 29–33. More recently DOs have updated these tenets based upon new scientific understanding from the fields of molecular biology and genetics. See Oliver Hayes and Philip Greenman, "Reprise on a Theme: Osteopathic Principles for the 21st Century," *The DO* 34 (December 1993): 21–28; Felix Rogers et al., "Proposed Tenets of Osteopathic Medicine and Principles for Patient Care," *JAOA* 102 (2002): 63–65.

21. Irvin Korr, "An Explication of Osteopathic Principles," in *Foundations for Osteopathic Medicine* 2nd ed., ed. Robert Ward (Philadelphia: Lippincott, Williams and Wilkins, 2002), pp. 12–18. See also Michael Seffinger et al., "Osteopathic Philosophy," in the same volume, pp. 3–12.

22. Jack Leahy, "Manipulation: A Survey of How DOs Feel About It," *OP* 38 (March 1972): 31–36.

23. National Center for Health Statistics, *Office Visits to Osteopathic Physicians January–December 1974: Provisional Data from the National Ambulatory Medical Care Survey* (Washington, D.C., n.d.), p. 19.

24. Lincoln Fry, "Preliminary Findings on the Use of Osteopathic Manipulative Treatment by Osteopathic Physicians," *JAOA* 96 (1996): 91–96; Shirley Johnson, Margot Kurtz, and J. C. Kurtz, "Variables Influencing the Use of Osteopathic Manipulative Treatment in Family Practice," *JAOA* 97 (1997): 80–87; Aguwa and Liechty, "Professional Identification and Affiliation," pp. 408–20.

25. Shirley Johnson and Margot Kurtz, "Diminished Use of Osteopathic Manipulative Treatment and Its Impact on the Uniqueness of the Osteopathic Profession," *Academic Medicine* 76 (2001): 821–28.

26. Singer, *2001 Annual Report*, p. 38

27. Robert Ward, ed. *Foundations for Osteopathic Medicine* (1st ed.) (Baltimore: Williams and Wilkins, 1996). (The second edition, published in 2002, is cited in note 21 of this chapter.).

28. For the problem of integrating distinctly osteopathic practices in clerkships and postdoctoral training see David Essig-Beatty et al., "Decline in Structural Examination Compliance in the Hospital Medical Record with Advancing Level of Training," *JAOA* 101 (2001): 501–8; Harry Friedman et al., "Effects of an Educational Intervention on Documentation of Palpatory and Structural Findings and Diagnosis," *JAOA* 96 (1996): 529; Jeffrey Bouley, "Osteopathic Medical Curricula Come Up Short on 'Osteopathy,'" *The DO* 41 (October 2000): 45–47; Donald Spaeth and Alfred Pheley, "Evaluation of Osteopathic Manipulative Treatment Training by Practicing Physicians in Ohio," *JAOA* 102 (2002): 145–50.

29. Jay Shubrook and John Dooley, "Effects of a Structured Curriculum in Os-

teopathic Manipulative Treatment (OMT) on Osteopathic Structural Examinations and Use of OMT for Hospitalized Patients," *JAOA* 100 (2000): 554-58; Warren Magnus and Russell Gamber, "Osteopathic Manipulative Treatment: Student Attitudes Before and After Intensive Clinical Exposure," *JAOA* 97 (1997): 109–13; Russell Gamber, Eric Gish, and Kathryn Herron, "Student Perceptions of Osteopathic Manipulative Treatment After Completing a Manipulative Medicine Rotation," *JAOA* 101 (2001): 395–400. One recently retired osteopathic physician and dean noted that a requirement for his graduation in 1959 included performing 500 osteopathic manipulative treatments in both his junior and senior years. No such requirement exists currently. See James Stookey, "Lessons from Mecca," *The DO* 41 (September 2000): 58–63.

30. John Licciardone and Kathryn Herron, "Characteristics, Satisfaction, and Perceptions of Patients Receiving Ambulatory Healthcare from Osteopathic Physicians: A Comparative National Survey," *JAOA* 101 (2001): 374–85. John Licciardone, Russell Gamber, and Kathryn Cardarelli, "Patient Satisfaction and Clinical Outcomes Associated with Osteopathic Manipulative Treatment," *JAOA* 102 (2002): 13–20; John Licciardone, Paul Brittain, and Samuel Coleridge, "Health Status and Satisfaction of Patients Receiving Ambulatory Care at Osteopathic Training Clinics," *JAOA* 102 (2002): 219–23.

31. For a review of neurophysiologic research by the Kirksville group in the 1960s and 1970s and their implications for osteopathic theory and practice see Irvin Korr, "The Spinal Cord as Organizer of Disease Processes; IV. Axonal Transport and Neurotrophic Function in Relation to Somatic Dysfunction," *JAOA* 80 (1981): 451–59.

32. See H. James Jones, "Somatic Dysfunction," in Ward, *Foundations for Osteopathic Medicine*, 2nd ed., pp. 1153–61.

33. Deborah Heath and Norman Gevitz, "The Research Status of Somatic Dysfunction," in Ward, *Foundations for Osteopathic Medicine*, 2nd ed., pp. 1188–93.

34. William Johnston, M. L. Elkiss, R. V. Marino, et al., "Passive Gross Motion Testing: Part II. A Study of Interexaminer Agreement," *JAOA* 81 (1982): 304–8; William Johnston, Myron Beal, G. Blum, et al. "Passive Gross Motion Testing: Part III. Examiner Agreement on Selected Subjects," *JAOA* 81 (1982): 309–13. Some researchers prefer the term *segmental dysfunction* to describe the phenomenon. See William Johnston, B. Allen, J. Hendra, et al., "Interexaminer Study of Palpation in Detecting Location of Spinal Segmental Dysfunction," *JAOA* 82 (1983): 839–45.

35. William Johnston, Joseph Vorro, and Robert Hubbard, "Clinical/Biomechanic Correlates for Cervical Function: Part I. A Kinematic Study," *JAOA* 85 (1985): 429–37; Joseph Vorro and William Johnston, "Clinical Biomechanic Correlates for Cervical Function: Part II. A Myoelectric Study," *JAOA* 87 (1987): 353–67; Joseph Vorro, William Johnston, and Robert Hubbard, "Clinical Biomechanic Correlates for Cervical Function: Part III. Intermittent Secondary Movements," *JAOA* 91 (1991): 145–55; Joseph Vorro and William Johnston, "Clinical Biomechanic Cervical Dysfunction: Part IV. Altered Regional Motor Behavior," *JAOA* 98 (1998): 317–23.

NOTES TO PAGES 183–184

36. William Johnston, James Hill, M. Elkiss, et al., "Identification of Stable So-matic Findings in Hypertensive Subjects by Trained Examiners Using Palpatory Examination," *JAOA* 81 (1982): 830–36; William Johnston, Albert Kelso, and Howard Babcock, "Changes in Presence of a Segmental Dysfunction Pattern As-sociated with Hypertension: Part 1. A Short-Term Longitudinal Study," *JAOA* 95 (1995): 243–55; William Johnston and Albert Kelso, "Changes in Presence of a Segmental Dysfunction Pattern Associated with Hypertension: Part II. A Long-Term Longitudinal Study," *JAOA* 95 (1995): 315–18; Myron Beal, "Palpatory Testing for Somatic Dysfunction in Patients with Cardiovascular Disease," *JAOA* 82 (1983): 822–31; John Cox, Sherman Gorbis, Lorane Dick, et al., "Palpable Musculoskeletal Findings in Coronary Artery Disease: Results of a Double-Blind Study," *JAOA* 82 (1983): 832–36; Myron Beal and George Kleiber, "Somatic Dys-function as a Predictor of Coronary Artery Disease," *JAOA* 85 (1985): 302–7; Alexander Nicholas, Domenic DeBias, Walter Ehrenfeuchter, et al., "A Somatic Component to Myocardial Infarction," *British Medical Journal* 291 (1985): 13–17. Myron Beal and John Morlock, "Somatic Dysfunction Associated with Pulmonary Disease," *JAOA* 84 (1984): 179–83; William Johnston, Albert Kelso, Donald Hollandsworth, et al., "Somatic Manifestations in Renal Disease: A Clinical Re-search Study," *JAOA* 87 (1987): 22–35; Jan Lei Iwata, J. Jerry Rodos, Thomas Glonek et al., "Comparing Psychotic and Affective Disorders by Musculoskeletal Structural Examination," *JAOA* 97 (1997): 715–20. For a study with negative re-sults and a critique of the methodologies of others see Robert Tarr, Richard Fee-ley, Daniel Richardson, et al., "A Controlled Study of Palpatory Diagnostic Pro-cedures: Assessment of Sensitivity and Specificity," *JAOA* 87 (1987): 296–301.

37. G. B. J. Andersson, T. Lucente, A. M. Davis, et al., "A Comparison of Os-teopathic Spinal Manipulation with Standard Care for Patients with Low Back Pain," *New England Journal of Medicine* 341 (1999): 1426–32. See also the accom-panying commentary by Joel Howell, "The Paradox of Osteopathy," pp. 1465–68. For the origins of this study see John Sprovieri, "AOA Launches Study of OMT, Back Pain," *The DO* 33 (September 1992): 60–61.

38. See *New England Journal of Medicine* 342 (2000): 817–20 for letters on the study and the accompanying commentary. See also Gilbert D'Alonzo, "What Is the Standard of Care for Patients with Low Back Pain," *JAOA* 99 (1999): 556; Jen-nifer Berger, "Top Journal Backs OMT," *The DO* 41 (February 2000): 28–33.

39. See Sandra Sleszynski and Albert Kelso, "Comparison of Thoracic Ma-nipulation with Incentive Spirometry in Preventing Postoperative Atelectasis," *JAOA* 93 (1993): 834–45; W. Randolph Purdy, Jesse Frank, and Brent Oliver, "Suboccipital Dermatomyotonic Stimulation and Digital Blood Flow," *JAOA* 96 (1996): 285–89; Kelly Jackson et al., "Effect of Lymphatic and Splenic Pump Techniques on the Antibody Response to Hepatitis B Vaccine: A Pilot Study," *JAOA* 98 (1998): 155–60; Michael Wells et al. "Standard Osteopathic Manipula-tive Treatment Acutely Improves Gait Performance in Patients with Parkinson's Disease," *JAOA* 99 (1999): 92–98; Donald Noll et al., "Benefits of Osteopathic Manipulative Treatment for Hospitalized Elderly Patients with Pneumonia," *JAOA* 100 (2000): 776–82; Russell Gamber et al., "Osteopathic Manipulative

Treatment in Conjunction with Medication Relieves Pain Associated with Fibromyalgia Syndrome: Results of a Randomized Clinical Pilot Project," *JAOA* 102 (2002): 321–25.

40. Kristie Aylett and Scott Stoll, "Introduction to the Texas College of Osteopathic Medicine at the University of North Texas Health Science Center at Ft. Worth," *American Academy of Osteopathy Journal* 11 (Winter 2001): 13–17; "The Profession's Emerging Research Center," *JAOA* 101 (2001): 323–24; Jacquie Goetz, "Rekindling Research," *The DO* 43 (April 2002): 22–25; idem, "Research Progress," *The DO* 43 (April 2002): 26–27. For other schools see idem, "Getting Off the Ground," *The DO* 43 (April 2002): 28–30.

41. One survey was conducted for the AOA by BSMG and Wirthlin Worldwide in early 1999. The other survey was conducted by the Gallup Organization for the AACOM. For published earlier data see Jay Lindquist, "Patient Confusion and Misperception About the Doctor of Osteopathy and the Medical Doctor," *Journal of Health Care Marketing* 8 (March 1988): 76–81; Charles Lamb, Ronald Haverstad, and Wade Lancaster, "Consumer Perceptions About Doctors of Osteopathy and Medical Doctors," *Journal of Health Care Marketing* 8 (December 1988): 53–57.

42. A number of DOs have been the subject of favorable local media coverage although it is rare for them to get national media attention for their accomplishments. When Elton Lehman DO of Ohio was named "country doctor of the year" by a Texas-based locum tenens firm he was featured in *USA Today, People,* and *Parade*. See "Ohio Doctor Named Country Doctor of the Year," *The DO* 39 (November 1998): 21–26. Sister Ann Brooks, an osteopathic physician as well as a nun, who practices in Mississippi in one of the poorest counties in the nation, has been featured on the television program *60 Minutes* as well as in national magazines and newspapers for her medical missionary work. See Marilyn Soltis, "Dr. Brooks Finds Reward in Serving God's Underserved," *The DO* 33 (February 1992): 102–08. It is also rare for the profession itself to get any coverage from national media. An exception is Abigail Zuger, "Scorned No More, Osteopathy on the Rise," *New York Times* (February 17, 1998): B9, B12.

43. Norman Gevitz, "Visible and Recognized: Osteopathic Invisibility Syndrome and the 2% Solution," *JAOA* 97 (1997): 168–70; Eckberg, "The Dilemma of Osteopathic Physicians," pp. 1111–20.

44. For media and public education efforts of the 1980s and early 1990s see for example Walter Wilson, "PR Report," *The DO* 22 (June 1982): 65–66; Al Boeck, "Department of Communications," *The DO* 26 (October 1985): 156–57; Donald Krpan, "Committee on Public Relations," *The DO* 33 (October 1992): 124–25; "Critics Give AOA Practice Marketing Kit Rave Reviews," *The DO* 33 (March 1992): 18 passim. Since 1980 the Ohio University College of Osteopathic Medicine has produced a weekday three-minute radio program called "Family Health" which is currently broadcast on about 300 radio stations across the country each day. It identifies its current host, Harold Thompson III, as either an osteopathic physician or a DO.

45. Howard Levine, "Would You Like the DO to Become a Household

Word?" *The DO* 33 (March 1992): 16. The national "unity campaign" had its origins in the activities of two state societies, Michigan and New Jersey, in the early 1990s. See Daniel Bade, "All for One," *The DO* 34 (January 1993): 70–79. For the national campaign see Jennifer Berger, "Getting the Word Out: Unity Campaign Promoters Profession to Public," *The DO* 40 (August 1999): 26 passim; Jacquie Goetz, "Creating the DO 'Brand,'" *The DO* 41 (November 2000): 28–33. The Tucson Osteopathic Medical Foundation has for years conducted an effective local public education campaign. See Jeffrey Bouley, "Promotion of DOs Blazing Away in Tucson," *The DO* 41 (October 2000): 34–38.

46. The extent of the media relations and public education problem is illustrated by what occurred in the December 2001 issue of *Self* magazine. The AOA had placed an advertisement on page 94. On page 54 of the same issue in a brief article entitled "When to Look for an MD Degree" the author stated quite remarkably, "If you have complicated health problems, stick with an MD. But if you're basically healthy, there's little reason not to see an osteopath; a DO has the same residency training as an MD." See "DOs Respond to Inaccurate Reporting," *The DO* 43 (March 2002): 19–20.

47. John Crosby, "The Road to Be Taken: AOA Leadership on Complementary and Alternative Medicine," *The DO* 43 (June 2001): 11–12.

48. It is beyond the scope of this book to discuss developments in osteopathy internationally. In the past two decades, interest in traditional osteopathic philosophy and methods has widened in Europe, Asia, Canada, Australia, New Zealand, and elsewhere. Since these international schools and practitioners neither teach nor embrace the full practice of medicine, DOs in this country have not yet agreed on what relations they should maintain with these groups. See John Sprovieri, "Academy Holds Global Forum on Osteopathic Practice," *The DO* 37 (June 1996): 43–45; Michael Fitzgerald, "AOA Trustees Create International Affairs Committee," *The DO* 37 (June 1996): 29–33; Lydia Hodges, "AOA Council Explores International Role for AOA," *The DO* 39 (May, 1998): 24–31.

49. George Thomas, "Bureau of State Government Affairs," *The DO* 39 (October 1998): 104; John Sprovieri, "Extenders or Independents," *The DO* 35 (August 1994): 68–71; Lydia Hodges, "Physicians Need to Forge New Relationships with Non-physician Clinicians," *The DO* 41 (June 2000): 66–69. See also Natalie Holt, "'Confusion's Masterpiece': The Development of the Physician Assistant Profession," *Bull Hist Med* 72 (1998): 246–78.

50. Christopher Durso, "Osteopathic As They Wanna Be," *The New Physician* 46 (May–June 1997): 16–22; Shirley Johnson and David Bordinat, "Professional Identity: Key to the Future of the Osteopathic Medical Profession in the United States," *JAOA* 98 (1998): 325–31; Katherine Miller, "The Evolution of Professional Identity: The Case of Osteopathic Medicine," *Social Science and Medicine* 47 (1998): 1739–48; Aguwa and Liechty, "Professional Identification and Affiliation," pp. 408–20. For an internal view of the prospects of osteopathic medicine see R. Michael Gallagher and Frederick Humphrey II, eds., *Osteopathic Medicine: A Reformation in Progress* (New York: Churchill Livingstone, 2001).

INDEX

Academy of Applied Osteopathy, 108–9

Accreditation Council for Graduate Medical Education, 165–70

Advertising, 41–42, 64–65, 188–89

Allopathy: definition of, 10; Still's opposition to, 25

American Association for the Advancement of Osteopathy, 54

American Association of Colleges of Osteopathic Medicine (*see also* Associated Colleges of Osteopathy), 178, 186, 187

American Association of Medical Colleges, 118, 134

American College of Surgeons, 92

American Hospital Association, 141

American Medical Association: acceptance of osteopathic members, 144–45; and California merger, 118, 128–34; efforts to achieve national merger, 142–46; on federal legislation and osteopathic medicine, 157; and homeopathy, 10, 48; inspection of osteopathic schools by, 125–28; and medical education, 85–87; opening postgraduate positions to DOs, 144–47; relations with American Osteopathic Association, 120–25, 178–80

American Medical Student Association, 174

American Osteopathic Association: and AMA inspection of osteopathic schools, 125–28; and California merger, 128–34; and California

Osteopathic Association, 118–19; code of ethics of, 63–65; combating imposters and imitators, 65–68; on federal legislation, 157–58; hospitals accredited by, 143, 162–64; licensure efforts of, 55–57, 82–84, 93–94, 98–100, 140–44, 178–79; membership in, 65, 179–80; origins of, 54; on osteopathic principles in hospitals, 146–47; and relations with AMA, 120–34, 135–46, 178; and research, 61, 63, 185–86; response of, to managed care, 160–61; Washington office of, 160

American Osteopathic Hospital Association, 161, 162

American School of Osteopathy (*see also* Kirksville College of Osteopathic Medicine): early classes of, 22–38; expanded curriculum of, 31–32; Flexner on, 88–90; formation of, 22–23; opposing three-year course, 59–60; relations of, with other schools, 48–53; surgery and obstetrics in, 70; vaccine and serum therapy in, 80

Appel, James Z., 125

Arizona College of Osteopathic Medicine, Midwestern University, 174, 177

Ashmore, Edythe, 60–61

Associated Colleges of Osteopathy, 57–59, 84

Atlantic School of Osteopathy, 48, 50, 59

53; displacing distinctive osteopathic procedures, 106–7; integration of, into osteopathic schools, 52–53, 75–84

Eccles, A. J., 34
Eclectic medicine, 10–11, 135
Eddy, Mary Baker, 14
Edward Via Virginia College of Osteopathic Medicine, 174, 177
Erlich, Paul, 76
Etherington, Frederick, 95, 124
Evans, A. L., 43, 44
Evans, Warren Felt, 15

Federation of State Medical Boards, 118, 178–79
Fishbein, Morris, 118
Flexner, Abraham, 86–90, 95
Foley, Alice, 110–11
Foraker, Joseph, 28
Fort Worth Osteopathic Hospital, 152
Franklin, Benjamin, 14
French Academy of Medicine, 14
French Academy of Sciences, 14
Furry, Frank, 78

Gaddis, Cyrus, 111
Gerard, Ralph Waldo, 104
Germ theory, 37–38, 76
Graham, Douglas, 34–35
Graham, Sylvester, 12–13
Greenwood, P. F., 29
Gregg, Alan, 104
Grunigen, Forest, 118, 119

Hahnemann, Samuel, 9–10
Hall, Marshall, 15
Hammond, William, 7
Hannah, F. W. and Mrs., 44
Harris, Wilfred, 58
Hartmann, George, 110
Hazzard, Charles, 32, 35, 36, 37
Health Care Financing Administration, 160
Health insurers (see also Medicaid; Medicare), 156, 159–60
Health maintenance organizations, 159
Health Manpower Act of 1968, 148

Health Professions Education Act of 1963, 148
Health systems agencies, 157, 158
Heatherington, J. Scott, 146
Helmer, George, 46–47
Henderson, Elmer, 120
Hildreth, Arthur, 56
Hill-Burton Act, 97, 111, 126
Hilton, John, 36
Hinckle, W. A., 77–78
Hippocrates, 33
Holmes, Oliver Wendall, Sr., 7, 10
Homeopathy, 9–10, 11, 25, 135
Hood, Wharton, 18
Hulett, C. M. T., 58, 74
Hulett, M. F., 111
Hydropathy, 12–13
Hypnotism, 14, 39

Imposters, 65–66
Independent practice associations, 159
Influenza pandemic of 1918–19, 81–82
International (foreign) medical graduates, 99, 150, 166

Jenner, Edward, 8
Joint Commission on the Accreditation of Hospitals (also called Joint Commission on the Accreditation of Health Care Facilities), 141
Journal of the American Medical Association, 118, 142, 149–50
Journal of the American Osteopathic Association, 60, 61, 75, 82, 185
Journal of Osteopathy, 26, 27, 43, 44, 47

Kansas City College of Osteopathy and Surgery (now University of Health Sciences–College of Osteopathic Medicine), 91, 92, 96, 97, 148–49, 177
Keesecker, Raymond, 113
Kentucky Board of Health, 45–46
Kinsman, J. Murray, 125
Kirksville College of Osteopathic Medicine (see also American School of Osteopathy), 90, 91, 92, 96, 97, 104–7, 108, 148–49, 177, 180